P9-DUO-795

PROGRESS IN BRAIN RESEARCH

VOLUME 36

BIOCHEMICAL AND PHARMACOLOGICAL MECHANISMS UNDERLYING BEHAVIOUR

PROGRESS IN BRAIN RESEARCH

PROGRESS IN BRAIN RESEARCH
VOLUME 36

BIOCHEMICAL AND PHARMACOLOGICAL MECHANISMS UNDERLYING BEHAVIOUR

EDITED BY

P. B. BRADLEY

Department of Pharmacology (Preclinical), The Medical School, University of Birmingham, Birmingham (England)

AND

R. W. BRIMBLECOMBE

Chemical Defence Establishment, Porton Down, Salisbury, Wiltshire (England)

ELSEVIER PUBLISHING COMPANY
AMSTERDAM / LONDON / NEW YORK
1972

ELSEVIER PUBLISHING COMPANY
335 JAN VAN GALENSTRAAT
P.O. BOX 211, AMSTERDAM, THE NETHERLANDS

AMERICAN ELSEVIER PUBLISHING COMPANY, INC.
52 VANDERBILT AVENUE, NEW YORK, N.Y. 10017

LIBRARY OF CONGRESS CARD NUMBER 72-190679

ISBN 0-444-40992-0

WITH 95 ILLUSTRATIONS AND 30 TABLES

PRINTED IN THE NETHERLANDS

List of Participants

ALDOUS, F. A. B., Chemical Defence Establishment, Porton Down, Salisbury (U.K.).
ANSELL, G. B., Department of Pharmacology (Preclinical), University of Birmingham, Birmingham (U.K.).
BALLANTYNE, B., Chemical Defence Establishment, Porton Down, Salisbury (U.K.).
BARSTAD, J. A. B., Department of Toxicology, Norwegian Defence Research Establishment, Kjeller (Norway).
BEBBINGTON, A., Chemical Defence Establishment, Porton Down, Salisbury (U.K.).
BERRY, W. K., Chemical Defence Establishment, Porton Down, Salisbury (U.K.).
BESWICK, F. W., Chemical Defence Establishment, Porton Down, Salisbury (U.K.).
BRADLEY, P. B., Department of Pharmacology (Preclinical), University of Birmingham, Birmingham (U.K.).
BRIGGS, I., Department of Pharmacology (Preclinical), University of Birmingham, Birmingham (U.K.).
BRIMBLECOMBE, R. W., Chemical Defence Establishment, Porton Down, Salisbury (U.K.).
BUXTON, D. A., Chemical Defence Establishment, Porton Down, Salisbury (U.K.).
CALLAWAY, S., Chemical Defence Establishment, Porton Down, Salisbury (U.K.).
COHEN, E. M., Medical Biological Laboratories, RVO-TNO & Department of Fundamental Pharmacology, University of Leiden, Leiden (Netherlands).
COOPER, G. H., Chemical Defence Establishment, Porton Down, Salisbury (U.K.).
COULT, D. B., Chemical Defence Establishment, Porton Down, Salisbury (U.K.).
COX, T., School of Pharmacy, University of Nottingham, Nottingham (U.K.).
CREASEY, N. H., Chemical Defence Establishment, Porton Down, Salisbury, (U.K.).
CROSSLAND, J., School of Pharmacy, University of Nottingham, Nottingham (U.K.).
DAVIES, J., School of Pharmacy, University of Bath, Bath, (U.K.).
FISHER, R. B., Department of Pharmacology (Preclinical), University of Birmingham, Birmingham, & CDE, Porton Down, Salisbury (U.K.).
FONNUM, F., Department of Toxicology, Norwegian Defence Research Establishment, Kjeller (Norway).
GORDON, J. J., Chemical Defence Establishment, Porton Down, Salisbury (U.K.).
GREEN, D. M., Chemical Defence Establishment, Porton Down, Salisbury (U.K.).
HANDS, D. H., Department of Pharmacology (Preclinical), University of Birmingham, Birmingham U.K.).
HEILBRONN, EDITH, Research Institute of National Defence, Sundbyberg (Sweden).
HOLLAND, P., Chemical Defence Establishment, Porton Down, Salisbury (U.K.).
HOLMES, R., Directorate of Biological & Chemical Defence, London (U.K.).
HOWELLS, D. J., Chemical Defence Establishment, Porton Down, Salisbury (U.K.).
HUGHES, ANNETTE, Chemical Defence Establishment, Porton Down, Salisbury (U.K.).
INCH, T. D., Chemical Defence Establishment, Porton Down, Salisbury (U.K.).
KEMP, K. H., Chemical Defence Establishment, Porton Down, Salisbury (U.K.).
KERKUT, G. A., Department of Physiology & Biochemistry, University of Southampton, Southampton (U.K.).
KING, A. R., Department of Pharmacology (Preclinical), University of Birmingham, Birmingham (U.K.).
KNIGHT, JOSEPHINE, Department of Pharmacology (Preclinical), University of Birmingham, Birmingham (U.K.).
LEADBEATER, L., Chemical Defence Establishment, Porton Down, Salisbury (U.K.).

MEETER, E., Medical Biological Laboratories, RVO-TNO, Rijswijk (Netherlands).
MOLENAAR, P. C., Department of Fundamental Pharmacology, University of Leiden, Leiden (Netherlands).
MOYLAN-JONES, R. J., Chemical Defence Establishment, Porton Down, Salisbury (U.K.).
MUIR, A. W., Chemical Defence Establishment, Porton Down, Salisbury (U.K.).
O'CONNOR, P. J., RAF Hospital, Wroughton (U.K.).
PATTLE, R. E., Chemical Defence Establishment, Porton Down, Salisbury (U.K.).
PINDER, R. M., Chemical Defence Establishment, Porton Down, Salisbury (U.K.).
POLAK, R. K., Medical Biological Laboratories, RVO-TNO, Rijswijk (Netherlands).
RAWLINS, J. S. P., Department of Director General Medical Services, Royal Navy, London (U.K.).
REDFERN, P., School of Pharmacy, University of Bath, Bath (U.K.).
RICK, J. T., Department of Psychology, University of Birmingham, Birmingham (U.K.).
RUTLAND, J. P., Chemical Defence Establishment, Porton Down, Salisbury (U.K.).
SAINSBURY, G., Chemical Defence Establishment, Porton Down, Salisbury (U.K.).
SAMUELS, GILLIAN M. R., Department of Pharmacology (Preclinical), University of Birmingham, Birmingham (U.K.).
SCHOCK, C., Chemical Defence Establishment, Porton Down, Salisbury (U.K.).
SINKINSON, D. V., Chemical Defence Establishment, Porton Down, Salisbury (U.K.).
SPANNER, SHEILA G., Department of Pharmacology (Preclinical), University of Birmingham, Birmingham (U.K.).
STORM-MATHISEN, J., Department of Toxicology, Norwegian Defence Research Establishment, Kjeller (Norway).
SWANSTON, D. W., Chemical Defence Establishment, Porton Down, Salisbury (U.K.).
SZERB, J. C., Department of Physiology & Biophysics, Dalhousie University, Halifax (Canada).
UPSHALL, D. G., Chemical Defence Establishment, Porton Down, Salisbury (U.K.).
VAN DER POEL, A. M., Department of Fundamental Pharmacology, University of Leiden, Leiden (Netherlands).
VINE, R. S., Home Office, Romney House, London (U.K.).
WALKER, R., Department of Physiology & Biochemistry, University of Southampton, Southampton (U.K.).
WATTS, P., Chemical Defence Establishment, Porton Down, Salisbury (U.K.).
WOODRUFFE, G. M., Department of Physiology & Biochemistry, University of Southampton, Southampton (U.K.).

List of Contributors

G. B. ANSELL, Department of Pharmacology (Preclinical), Medical School, Birmingham B15 2TJ, England.

B. C. BARRASS, Chemical Defence Establishment, Porton Down, Salisbury, Wiltshire, England.

J. A. B. BARSTAD, Norwegian Defence Research Establishment, Division of Toxicology, P.O. Box 25, Kjeller, Norway.

P. B. BRADLEY, Department of Pharmacology (Preclinical), Medical School, Birmingham B15 2TJ, England.

R. W. BRIMBLECOMBE, Medical Division, Chemical Defence Establishment, Porton Down, Salisbury, Wiltshire, England.

D. A. BUXTON, Chemical Defence Establishment, Porton Down, Salisbury, Wiltshire, England.

J. A. DAVIES, School of Pharmacy, University of Bath, Bath, England.

F. FONNUM, Norwegian Defence Research Establishment, Division for Toxicology, P.O. Box 25, Kjeller, Norway.

D. M. GREEN, Medical Division, Chemical Defence Establishment, Porton Down, Salisbury, Wiltshire, England.

E. HEILBRONN, Research Institute of the Swedish National Defence, Avdelning 1, Box 416, S-172 04 Sundbyberg 4, Sweden.

T. D. INCH, Chemical Defence Establishment, Porton Down, Salisbury, Wiltshire, England.

G. A. KERKUT, Department of Physiology and Biochemistry, University of Southampton, Southampton, England.

E. MEETER, Medical Biological Laboratories, RVO-TNO, Lange Kleiweg 139, Rijswijk ZH, Netherlands.

J. T. RICK, Department of Psychology, University of Birmingham, Birmingham B15 2TT, England.

G. M. R. SAMUELS, Tunstall Laboratory, Shell Research, Sittingbourne, Kent, England.

J. STORM-MATHISEN, Norwegian Defence Research Establishment, Division for Toxicology, P.O. Box 25, Kjeller, Norway.

J. C. SZERB, Department of Physiology and Biophysics, Dalhousie University, Sir Charles Tupper Medical Building, Halifax, Nova Scotia, Canada.

A. M. VAN DER POEL, Department of Fundamental Pharmacology, University of Leiden, Wassenaarseweg 62, Leiden, Netherlands.

Preface

The papers contained in this Volume were presented at a meeting held at the Chemical Defence Establishment, Porton Down, Salisbury, on March 22 and 23, 1971, and which was attended by government scientists from the U.K., Norway, Sweden and the Netherlands, together with a number of academic research workers.

While there exist in many countries brain research institutes where neurobiologists from various disciplines work side by side and have daily contact for the exchange of ideas and experimental findings, in the U.K. such research is fragmented. Thus, certain individuals working in departments of anatomy, physiology, pharmacology and psychology are engaged upon investigations into brain function in their own disciplines but there is no coordinated effort, nor are there adequate opportunities for liaison between different disciplines. One way of attempting to overcome this isolation of brain research workers is by meetings or symposia and one of the purposes of the C.D.E. Symposium was to achieve this end.

Additionally, the meeting served the purpose of bringing together scientists in government research laboratories and academic research workers who do not often have such opportunities for exchange of ideas, etc. It is to be hoped that further meetings will be held in the future along similar lines but on different topics.

As the Symposium had to be limited in the number of participants attending, it was decided to concentrate in this first meeting on two main aspects of brain research, namely the study of biochemical and pharmacological mechanisms. In particular, since these two approaches tend to be pursued independently, it was hoped that a greater degree of integration between biochemistry and pharmacology might ensue and that their relevance in terms of behaviour become more apparent. The first day was therefore devoted to papers dealing in the main with biochemical mechanisms and on the second day papers on the actions of drugs producing changes in behaviour were presented. The fact that more than half the papers presented were concerned to a greater or lesser extent with cholinergic mechanisms in the central nervous system reflects, in our opinion, the relative importance of acetylcholine in brain function, although of late this transmitter has been somewhat neglected in favour of others.

We are grateful to the Director of C.D.E., Mr. G. N. Gadsby, for making facilities available for this Symposium and to the staff of the Establishment for their assistance with the organization. Thanks are also due to Miss Sally Clements for helping to edit these proceedings and to Miss Josephine Knight for her excellent work during the meeting in keeping track of the discussion and discussants.

P. B. Bradley
R. W. Brimblecombe

Contents

The Application of Zonal Centrifugation to the Study of Some Brain Subcellular Fractions

G. B. ANSELL AND SHEILA SPANNER

Department of Pharmacology (Preclinical), The Medical School, Birmingham, B15 2TJ (Great Britain)

The first attempt to separate subcellular fractions from brain tissue was by Brody and Bain (1952) but, with the benefit of hindsight, it is clear that the fractions they obtained were extremely heterogeneous. The major impetus for improving the techniques of subcellular fractionation stemmed from the observation of Hebb and Smallman (1956) that a high proportion of the choline acetyltransferase (EC 2.1.3.6) in brain tissue could be located in the mitochondrial fraction when this was prepared by the method of Brody and Bain (1952). Hebb and Whittaker (1958) then demonstrated that acetylcholine (ACh) was also associated with the crude mitochondrial fraction and made the important observation that the ACh-containing particles could also be separated from the mitochondria. This led to an intensive investigation by Whittaker's group and by De Robertis and his colleagues of which an excellent account is given by Whittaker (1965). They succeeded, using different density gradients, in characterising the ACh-containing organelles as the pinched-off nerve terminals ("synaptosomes") which are formed from nerve endings when brain is homogenised in isotonic sucrose.

A sophisticated technique was eventually developed using a combination of rate centrifugation in 0.32 M-sucrose and isopycnic centrifugation using sucrose gradients of between 0.32 M and 1.2 M. The P_2, or crude mitochondrial fraction, sediments in 0.32 M-sucrose between 5×10^4 and 3.6×10^5 *g*/min. The subcellular components of this fraction were separated by Whittaker (1965) into myelin, synaptosomes and mitochondria on a discontinuous gradient of 0.32 M, 0.8 M and 1.2 M-sucrose in a tube and centrifuged at 3×10^6 *g*/min. Under these conditions the myelin floated between the 0.32 M and 0.8 M-sucrose, the mitochondria formed a pellet at the bottom of the tube and the synaptosomes settled in a diffuse band between 0.8 M and 1.2 M-sucrose. Subsequently it was observed that these fractions showed considerable heterogeneity when the P_2 fraction was subjected to a more refined sucrose gradient (Whittaker, 1968). It was also found that the mitochondrial fraction contained the lysosomes (Koenig *et al.*, 1964).

The small amount of material obtained and the relative difficulty of obtaining the synaptosomal fraction from the middle of the gradient, prompted us to apply zonal centrifugation to the separation of the P_2 fraction.

The zonal centrifugation of the P_2 fraction

The technique of zonal centrifugation was developed by Anderson but there have, as yet, been few applications of this technique to the separation of the subcellular components of brain tissue (Barker *et al.*, 1970; Cotman *et al.*, 1968; Kornguth *et al.*, 1971; Mahler *et al.*, 1970; Rodnight *et al.*, 1969; Shapira *et al.*, 1970; Spanner and Ansell, 1970). With the exception of Rodnight *et al.* (1969) and Spanner and Ansell (1970), these workers have used continuous density gradients of sucrose or caesium chloride or a Ficoll–sucrose mixture, but, in our hands, this did not produce well-defined peaks when the P_2 fraction was subjected to zonal centrifugation (Spanner and Ansell, 1971). The use of a shallow, discontinuous sucrose gradient gives a much better separation with apparently well-defined and discrete peaks. Essentially it was shown that, after the removal of the myelin by a preliminary separation of the P_2 fraction in 0.8 M-sucrose in tubes (Fig. 1), the remainder of the fraction could be separated into 6 protein-containing peaks on a suitable gradient (Fig. 2). These peaks have been examined and partially identified by means of the electron microscope and enzyme markers.

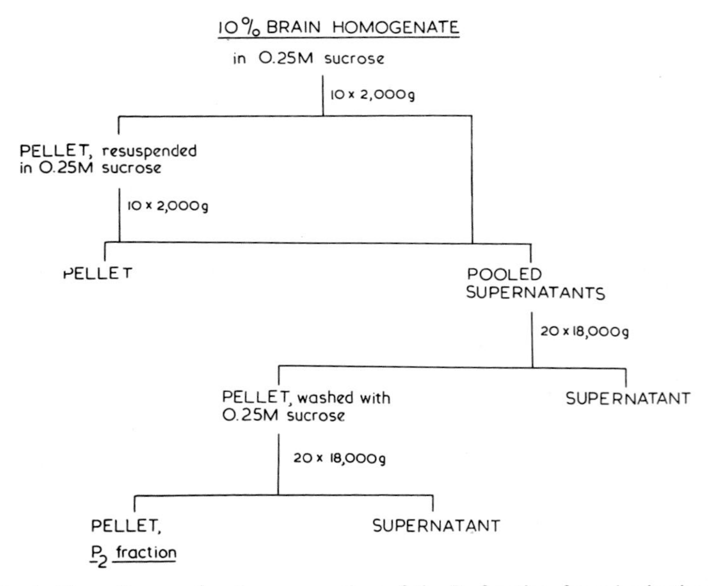

Fig. 1. Flow diagram for the preparation of the P_2 fraction from brain tissue.

Some features of Fig. 2 warrant attention. It has been established from work with tubes that mitochondria form the major part of the fraction sedimenting in 1.4 M-sucrose and this is borne out in zonal studies by the high level of succinic dehydrogenase (succinate: (acceptor) oxidoreductase, EC 1.3.99.1) (Table I). There was a

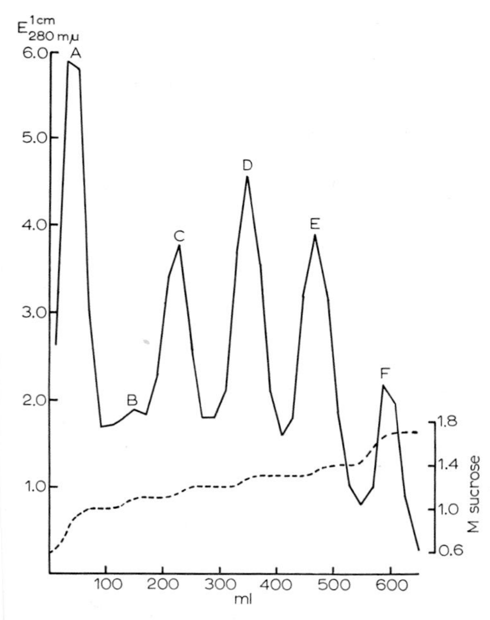

Fig. 2. The subfractionation of the P_2 fraction from rabbit cortex on a discontinuous sucrose gradient in a zonal rotor after the initial removal of myelin. The BXIV rotor, capacity 650 ml, was used and spun for 8×10^6 *g* min. (———) $E^{1\,cm}_{280\,m\mu}$ (not a quantitative protein estimation); (-----) sucrose concentration (M). For information about the individual peaks, see text.

TABLE 1

THE DISTRIBUTION OF ENZYME MARKERS AMONG THE COMPONENTS OF THE P_2 FRACTION SHOWN IN FIG. 2.

	A	*B*	*C*	*D*	*E*	*F*
Acetylcholinesterase	45	17	22	10	4	2
Occluded lactic dehydrogenase	0	3	49	34	10	4
Succinic dehydrogenase	2	4	15	33	43	2
β-Glucuronidase	5	5	12	14	12	52

Values are expressed as a percentage of that found in the whole P_2 fraction.

clear separation of mitochondria from a fraction free from succinic dehydrogenase activity and sedimenting between 1.4 M and 1.7 M-sucrose. This fraction had a high level of β-glucuronidase (β-D-glucuronide glucuronohydrolase, EC 3.2.1.31) activity and acid phosphatase (orthophosphoric monoester phosphohydrolase, EC 3.1.3.2) activity, two lysosomal enzyme markers (Table I). It can be seen from Table I, for example, that 50% of the β-glucuronidase activity was found in this fraction. After treating the fractions E and F with acridine orange, a clear distinction could be seen with the fluorescence microscope between the bright orange fluorescence of the lysosomal fraction and the bright green of the mitochondria (*cf.* Koenig, 1969).

This centrifugation method provides a quicker means of obtaining relatively large quantities of brain lysosomes than those reported in the paper of Sellinger and Nordrum (1969).

There were two clearly separated peaks, C and D, sedimenting at 1.2 M and 1.3 M-sucrose (Fig. 2). Examination by the electron microscope showed them both to have the morphological characteristics of synaptosomes. It is well established that a good enzyme marker for intact synaptosomes is occluded lactic dehydrogenase (L-lactate: NAD oxidoreductase, EC 1.1.1.27) (Marchbanks, 1967), a component of the cell sap. As can be seen in Table I, 83% of the occluded form of the enzyme of the original P_2 fraction is shared between peaks C and D. Both also contained succinic dehydrogenase activity owing to the presence of intraterminal mitochondria. The membrane marker acetylcholinesterase (acetylcholine hydrolase, EC 3.1.1.7) was also present in these peaks and was notably absent from the mitochondrial and lysosomal fractions (Table I).

The more diffuse peak which spread throughout the 0.8 M-sucrose band had the characteristics of plasma membranes in that there was a high acetylcholinesterase and 5′-nucleotidase (5′-ribonucleotide phosphohydrolase, EC 3.1.3.5) activity. Using the intact P_2 fraction, complete with myelin, as starting material and especially tailored gradients, the myelin and plasma membranes can be separated from each other and from the lighter synaptosomal peak.

The two synaptosomal peaks

The explanation for the two synaptosomal peaks obtained in these experiments is not known for certain. It may be that this separation is a function of size but it is not an artefact of the stepwise gradient as can be seen from the definite "shoulders" on the peak obtained when the P_2 fraction is subjected to a continuous gradient (Spanner and Ansell, 1971). Other workers have demonstrated at least two synaptosomal populations (*e.g.*, De Robertis, 1967; Lemkey-Johnston and Dekirmenjian, 1970; Whittaker, 1968). Interest naturally lies in attempts to demonstrate different transmitter content or a capacity for the differential uptake and metabolism of chemical transmitters. Iversen and Snyder (1968) have shown in their separations that there appeared to be at least two synaptosomal populations in the striatum, one of which could accumulate labelled γ-aminobutyric acid and the other denser population which could accumulate labelled noradrenaline. Hökfelt *et al.* (1970) incubated the synaptosomal fraction obtained from the hypothalamus and nucleus caudatus putamen with α-methylnoradrenaline in Ringer–bicarbonate solution and showed by electron microscopy that certain synaptosomes, probably those containing small granular vesicles, were able to take up the monoamine preferentially. Very recently Kuhar *et al.* (1971) have succeeded in obtaining a partial separation of synaptosomes accumulating γ-aminobutyric acid, 5-hydroxytryptamine and noradrenaline.

(1) Differential uptake of α-methylnoradrenaline

To see if synaptosomal peaks C and D could be differentiated by a similar method, we adapted the techniques of Iversen and Snyder (1968) and of Hökfelt *et al.* (1970) as follows. Adult female rats were given the monoamine oxidase inhibitor pargyline in a dose of 250 mg/kg body wt. and killed after 16 h. The brains were fractionated and the synaptosomal peaks C and D obtained by zonal centrifugation as already described. The sucrose concentration in these peaks was carefully reduced and the fractions centrifuged in tubes to bring down representative pellets. Each pellet was then suspended in Krebs bicarbonate Ringer and centrifuged for 20 min at 15,000 $\times$ *g*. Each pellet was re-suspended in 2 ml of bicarbonate Ringer containing 0.4 mg of ascorbic acid and 20 μg of α-methylnoradenaline and incubated for 30 min at 37°C in the presence of 95% O_2 and 5% CO_2. Ice-cold bicarbonate Ringer (8 ml) was added and the synaptosomal pellets again obtained by centrifugation. Smears were made of the pellets on glass slides and dried *in vacuo* over P_2O_5. After exposure to formaldehyde gas, the smears were mounted in Entellon and examined by fluorescence microscopy.

There was a marked qualitative difference between the two synaptosomal peaks in that peak D showed a significant green fluorescence almost completely absent from the other. From the present study it does appear that peak D is enriched in synaptosomes from monoamine-containing neurones though further quantitative studies (*e.g.*, the uptake of different radioactive amines and the determination of monoamineoxidase) are required to substantiate these findings.

(2) Uptake of [Me-^{14}C]choline in vivo

Chakrin and Whittaker (1969) demonstrated that intracerebrally injected or topically applied [*Me*-^{3}H]choline was rapidly distributed throughout the brain and readily labelled the "labile bound" and, to a lesser extent, the "stable bound" (vesicular) ACh. Ansell and Spanner (1968) showed that intracerebrally injected [*Me*-^{14}C]-choline was also rapidly phosphorylated and incorporated into a lipid-bound form in whole brain tissue. The rapid utilization of free choline after intracerebral injection contrasts with the more recent finding of Ansell and Spanner (1971) that there is no measurable transport of free choline to the brain from the blood *in vivo* and that the organ may well receive its supply of choline, and hence the choline for ACh-synthesis, in a lipid-bound form from the blood.

In some preliminary studies, the uptake of [*Me*-^{14}C]choline into the subcellular fractions of brain has been measured. Rats were injected intracerebrally with 1 μCi (0.018 μmoles) of [*Me*-^{14}C]choline, and after 5 h the animals were killed and the brains were homogenized to obtain the subcellular fractions. Fig. 3 demonstrates that the largest percentage incorporated was into the P_2 fraction, and, of this fraction, nearly 50% was found in the plasma membranes. The uptake into synaptosomes was about 12% of the uptake into the P_2 fraction. There was no significant difference between the two synaptosomal peaks in this experiment. It seems clear that there is an active transport of free choline into isolated synaptosomes *in vitro* (Diamond and Kennedy, 1969; Marchbanks, 1968) but only a small amount of the incorporated choline is converted to ACh. *In vivo* we found that 40% of the labelled choline in the synaptosomes was in a lipid-bound form 5 h after injection, and work is in progress

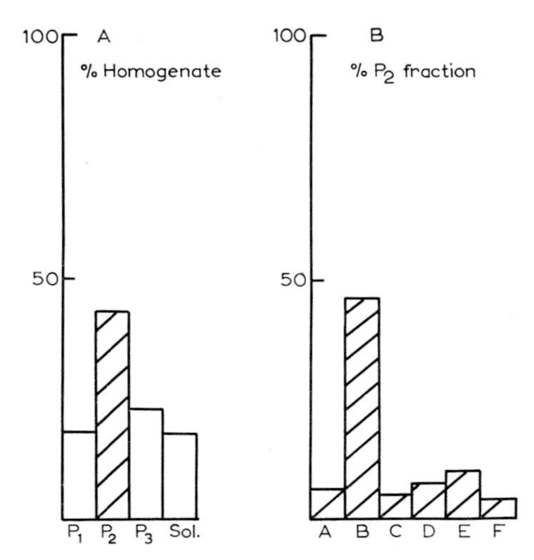

Fig. 3. The uptake of [*Me*-^{14}C]choline into the subcellular fractions of whole rat brain 5 h after intracerebral injection. A, The distribution of radioactivity as a percentage of that in the homogenate in the nuclei + debris (P_1), crude mitochondrial fraction (P_2), microsomal fraction (P_3) and soluble cell sap (Sol.); and B, The distribution of radioactivity in the components of the P_2 fraction. Peaks are the same as those described in Fig. 2.

on the zonal separation of the synaptosomal components to establish the distribution of the various choline compounds and their radioactivity. The situation is complicated by the presence of phospholipases since they are capable of liberating water-soluble choline compounds from phosphatidylcholine which is a significant component of the synaptosomal membrane. The activity of these enzymes is being studied so that an overall picture of the metabolism of choline within the brain with special reference to synaptosomes can be obtained.

The subcellular fractionation of brain tissue has developed significantly over the past decade and can be applied to discrete parts of the CNS. It is increasingly possible to prepare more homogeneous synaptosomal populations and it is likely that zonal centrifugation will make such populations more readily available and in larger amounts.

SUMMARY

The separation of the components of the myelin-free crude mitochondrial fraction of whole brain tissue in centrifuge tubes is compared with a separation by zonal centrifugation. On a shallow, step-wise gradient of 0.8–1.7 M sucrose in a BXIV rotor of 650 ml capacity, it was possible to obtain lysosomal, mitochondrial, synaptosomal and plasma membrane fractions after spinning for 2 h at 67,000 × *g*. These fractions were characterised by enzyme markers and other means. At least two synaptosomal populations could be clearly separated, one of which could actively take up α-methylnoradrenaline. Some preliminary studies on the uptake of [*Me*-^{14}C]-choline into sub-cellular components after intracerebral injection are also described.

ACKNOWLEDGEMENTS

We are grateful to the Multiple Sclerosis Society of Great Britain for funds with which the zonal rotor and the labelled choline were purchased. We would also like to thank Mr. J. Candy for examining some fractions by fluorescence microscopy and Professor P. B. Bradley for his interest in the work.

REFERENCES

ANSELL, G. B. AND SPANNER, S. (1968) The metabolism of [*Me*-^{14}C]choline in the brain of the rat *in vivo*. *Biochem. J.*, **110**, 201–206.

ANSELL, G. B. AND SPANNER, S. (1971) Studies on the origin of choline in the brain of the rat. *Biochem. J.*, **122**, 741–750.

BARKER, L. A., DOWDALL, M. J., ESSMAN, W. B. AND WHITTAKER, V. P. (1970) The compartmentation of acetylcholine in cholinergic nerve terminals. In *Drugs and Cholinergic Mechanisms in the CNS*, E. HEILBRONN and A. WINTER (Eds.), Res. Inst. of Natl. Def., Stockholm. Almqvist and Wiksell, Stockholm, pp. 193–223.

BRODY, T. M. AND BAIN, J. A. (1952) A mitochondrial preparation from mammalian brain. *J. biol. Chem.*, **195**, 685–696.

CHAKRIN, L. W. AND WHITTAKER, V. P. (1969) The subcellular distribution of [*N-Me*-^{3}H]acetylcholine synthesised by brain *in vivo*. *Biochem. J.*, **113**, 97–107.

COTMAN, C., MAHLER, H. R. AND ANDERSON, N. G. (1968) Isolation of a membrane fraction enriched in nerve-end membranes from rat brain by zonal centrifugation. *Biochim. biophys. Acta (Amst.)*, **163**, 272–275.

DE ROBERTIS, E. (1967) Ultrastructure and cytochemistry of the synaptic region. *Science*, **156**, 907–914.

DIAMOND, I. AND KENNEDY, E. P. (1969) Carrier-mediated transport of choline into synaptic nerve endings. *J. biol. Chem.*, **244**, 3258–3263.

HEBB, C. O. AND SMALLMAN, B. N. (1956) Intracellular distribution of choline acetylase. *J. Physiol. (Lond.)*, **134**, 385–392.

HEBB, C. O. AND WHITTAKER, V. P. (1958) Intracellular distributions of acetylcholine and choline acetylase. *J. Physiol. (Lond.)*, **142**, 187–196.

HÖKFELT, T., JONSSON, G. AND LIDBRINK, P. (1970) Electron microscopic identification of monoamine nerve ending particles in rat brain homogenates. *Brain Res.*, **22**, 147–151.

IVERSEN, L. L. AND SNYDER, S. H. (1968) Synaptosomes: different populations storing catecholamines and gamma-aminobutyric acid in homogenates of rat brain. *Nature (Lond.)*, **220**, 796–798.

KOENIG, H. (1969) Lysosomes. In *Handbook of Neurochemistry II. Structural Neurochemistry*, A. LAJTHA (Ed.), Plenum Press, London, pp. 255–301.

KOENIG, H., GAINES, D., MCDONALD, T., GRAY, R. AND SCOTT, J. (1964) Studies of brain lysosomes. 1. Subcellular distribution of five acid hydrolases, succinate dehydrogenase and gangliosides in rat brain. *J. Neurochem.*, **11**, 729–743.

KORNGUTH, S. E., FLANGAS, A. L., SIEGEL, F. L., GEISON, R. L., O'BRIEN, J. F., LAMAR, JR., C. AND SCOTT, G. (1971) Chemical and metabolic characteristics of synaptic complexes from brain isolated by zonal centrifugation in a cesium chloride gradient. *J. biol. Chem.*, **246**, 1177–1184.

KUHAR M. J., SHASKAN, E. G. AND SNYDER, S. H. (1971) The subcellular distribution of endogenous and exogenous serotonin in brain tissue: comparison of synaptosomes storing serotonin, norepinephrine and γ-aminobutyric acid. *J. Neurochem.*, **18**, 333–343.

LEMKEY-JOHNSTON, N. AND DEKIRMENJIAN, H. (1970) The identification of fractions enriched in non-myelinated axons from rat whole brain. *Exp. Brain Res.*, **11**, 392–410.

MAHLER, H. R., MCBRIDE, W. AND MOORE, W. J. (1970) Isolation and characterisation of membranes from rat cerebral cortex. In *Drugs and Cholinergic Mechanisms in the CNS*, E. HEILBRONN AND A. WINTER (Eds.), Res. Inst. of Natl. Def., Stockholm. Almqvist and Wiksell, Stockholm, pp. 225–244.

MARCHBANKS, R. M. (1967) The osmotically sensitive potassium and sodium compartments of synaptosomes. *Biochem. J.*, **104**, 148–157.

MARCHBANKS, R. M. (1968) The uptake of [^{14}C] choline into synaptosomes *in vitro*. *Biochem. J.*, **110**, 533–541.

RODNIGHT, R., WELLER, M. AND GOLDFARB, P. S. G. (1969) Large scale preparation of a crude membrane fraction from ox brain. *J. Neurochem.*, **16**, 1591–1597.

SELLINGER, O. Z. AND NORDRUM, L. M. (1969) A regional study of some osmotic, ionic and age factors affecting the stability of cerebral lysosomes. *J. Neurochem.*, **16**, 1219–1229.

SHAPIRA, R., BINKLEY, F., KIBLER, R. F. AND WUNDRAM, I. J. (1970) Preparation of purified myelin of rabbit brain by sedimentation in a continuous sucrose gradient. *Proc. Soc. exp. Biol. (N.Y.)*, **133**, 238–245.

SPANNER, S. AND ANSELL, G. B. (1970) The use of zonal centrifugation in the preparation of subcellular fractions from brain tissue. *Biochem. J.*, **119**, 45P.

SPANNER, S. AND ANSELL, G. B. (1971) Preparation of subcellular fractions from brain tissue. In *Separations with Zonal Rotors*, E. REID (Ed.), Wolfson Bioanalytical Centre of the University of Surrey, Guildford, pp. V-3.1–3.7.

WHITTAKER, V. P. (1965) The application of subcellular fractionation techniques to the study of brain function. *Progr. Biophys. molec. Biol.*, **15**, 39–96.

WHITTAKER, V. P. (1968) The morphology of fractions of rat forebrain synaptosomes separated on continuous sucrose density gradients. *Biochem. J.*, **106**, 412–417.

DISCUSSION

KERKUT: Do you have any evidence as to whether labelled choline is specifically accumulated in the C-fraction?

ANSELL: No, we haven't, in that preliminary experiments have not shown this. The problem of choline uptake, we find, is very much more complex than we thought originally and we are not now sure whether we should be doing experiments on the actual uptake of *free* choline. Our other work is showing that the brain does not make choline, but has to bring it from outside, probably as either phosphatidylcholine or as lysophosphatidylcholine, which pass through the blood-brain barrier, then yield free choline. Therefore, although interesting results can be obtained after injection of labelled choline, we get slightly worried about the relevance of the uptake of free choline into synaptosomes. This is why we want to fractionate the synaptosomes to see which fractions, if any, are capable of yielding free choline under these conditions.

CREASEY: Zonal centrifugation has the advantage of being able to fractionate large quantities of brain tissue; on the other hand, behavioural changes may be caused by changes in relatively small and localized regions of the brain; so how are the techniques you described relevant to behavioural changes?

ANSELL: I am not the first biochemist who has had to defend himself in such a situation. It is true that as described at this meeting the method is unsophisticated in that we have applied it to large pieces of tissue. When I said that it was a "bulk method" I meant that one would hope to use a large amount of material to obtain small amounts of rather more specific fractions. There is no reason, however, why the hippocampus, for example, or some other area should not be used. Several of these could be pooled and subjected to bulk fractionation to study particular nerve endings. I don't think I have stated that zonal fractionation is going to solve behavioural problems. What I hope it is going to tell us is something about the biochemistry of different types of nerve endings with a view to finding, for example, which transmitters are present in them.

FONNUM: Due to the morphological heterogeneity of nerve terminals, one cannot really expect to separate synaptosomes containing different transmitters from whole brain. The chances of success will be greatly enhanced by working on regions of the brain. Nafstad and Blackstad (1966) have shown that axosomatic nerve terminals in the hippocampus contain a higher proportion of mitochondria than axodendritic nerve terminals, and separation of these terminals should therefore be possible. But this difference does not necessarily hold true for any other part of the brain.

ANSELL: I take your point, Dr. Fonnum. We have been trying to get the zonal method going and now I think we can do more with it and look for more specialized areas. What we cannot quite understand in this methodology, and I discussed this with Dr. Mahler at a meeting in Skokloster, is that he holds that a continuous gradient is considerably better than a discontinuous gradient. We have found that with shallow steps far better defined fractions are obtained. I am not, of course, saying that fractions C and D could not be subsequently sub-fractionated, but we would like to find out first why two such apparently discrete fractions should exist.

HEILBRONN: Could you elaborate on the phospholipase content of lysosomes?

ANSELL: Yes; the lysosomes that we have obtained from brain tissue certainly do not contain any phospholipases that can attack ethanolamine phospholipids. I would not like to say at this point in the investigation that they do not contain phospholipases attacking choline lipids.

REFERENCES

NAFSTAD, P. H. J. AND BLACKSTAD, T. W. (1966) Distribution of mitochondria in pyramidal cells and boutons in hippocampal cortex. *Z. Zellforsch.*, **73**, 234–245.

Molecular Properties of Choline Acetyltransferase and Their Importance for the Compartmentation of Acetylcholine Synthesis

F. FONNUM AND D. MALTHE-SØRENSSEN

Norwegian Defence Research Establishment, Division for Toxicology, 2007 Kjeller (Norway)

Choline acetyltransferase (ChAc) is responsible for the synthesis of the chemical transmitter acetylcholine (ACh) and is therefore an important enzyme in nervous tissue. There is an excellent correlation between the level of ACh and the level of ChAc in different parts of the nervous system (Silver, 1967), indicating that this enzyme governs the level of ACh. The localization of ChAc is therefore synonymous with the site of synthesis of ACh and knowledge of this site gives important information as to the further processes necessary for the uptake and storage of the neurotransmitter.

Distribution of ChAc in the neurone

It is generally accepted that ChAc is recovered in both a soluble and a particulate form after homogenization of brain tissue in iso-osmotic sucrose (Hebb and Smallman, 1956; Hebb and Whittaker, 1958). The soluble form of ChAc is derived by disruption of cholinergic cell bodies and probably also axons and dendrites, whereas the particulate form is obtained from the detached nerve terminals (synaptosomes).

TABLE 1

PROPORTION OF ChAc AND LACTATE DEHYDROGENASE (LDH) IN HIGH SPEED SUPERNATANT FROM SUCROSE PHOSPHATE BUFFER HOMOGENATE.

	% Soluble	
	ChAc	*LDH*
Cat ventral root	81	85
Rat ventral root	76	79
Rat sciatic nerve	80	83
Rat phrenic nerve	72	69
Cat ventral horn	58	77
Cat nucleus ruber	45	76
Rat cerebral cortex	28	51
Cat nucleus interpositus	28	82
Rat hippocampus	21	48
Rat caudate nucleus	12	70

The tissue was homogenized in 0.32 M sucrose and centrifuged in 0.30 mmoles sodium phosphate buffer + 0.25 mmoles sucrose at 17000 × *g* for 60 min.

The particulate ChAc is present in an occluded form and full enzyme activity is only obtained after treatment with agents that disrupt membranes such as organic solvents (Hebb and Smallman, 1956) or detergents (Fonnum, 1966a). The distribution of ACh parallels that of ChAc in these fractions.

The relative distribution of ChAc between the soluble and particulate form therefore provides information as to the presence of cholinergic cell bodies or nerve terminals in a certain region (Table I). The main part of ChAc from the hippocampus and caudate nucleus is obtained in particulate form indicating that in these tissues the cholinergic structures are mainly present as nerve terminals. In the cerebellar nuclei and cerebral cortex there is slightly more soluble ChAc, whereas in the ventral horn and red nucleus a relatively larger proportion of the enzyme is soluble, indicating that in these regions a comparatively high proportion of the cholinergic structures are cell bodies. In the ventral root, sciatic nerve and phrenic nerve most of the enzyme is obtained in a soluble form, indicating that in peripheral axons the enzyme is largely soluble. Data on the proportion of soluble lactate dehydrogenase, a general cytoplasmic marker, in the different fractions are also given.

Compartmentation of ChAc within the synaptosome

Methods have been developed by Whittaker (1965) and De Robertis (1967) for studying the compartmentations of ACh, ChAc and acetylcholinesterase (AChE) within the synaptosome. By re-suspending the synaptosomes in water, it was found that they burst, and that the different constituents could be separated by differential (De Robertis *et al.*, 1963) or density gradient centrifugation (Whittaker *et al.*, 1964). Both techniques led to the conclusion that ACh was present both in cytoplasma and in synaptic vesicles, whereas AChE was bound to membranes. The two groups disagreed, however, with regard to the localization of ChAc. De Robertis *et al.* (1963) maintained that the enzyme in the rat was strongly attached to the synaptic vesicles whereas Whittaker *et al.* (1964) showed that the enzyme in the guinea pig was obtained from the cytoplasma.

Subsequent work by McCaman *et al.* (1965), Tuček (1966a, b) and Fonnum (1966b, 1967) showed that there were considerable species differences with regard to the proportion of particulate ChAc obtained from disrupted synaptosomes. The enzyme was obtained mainly in the soluble form from pigeon synaptosomes, in partly soluble form from guinea-pig synaptosomes and in largely particulate form from rat, rabbit and cat synaptosomes (Fig. 1). The particulate forms of ACh and ChAc from the disrupted synaptosomes of rat brain behaved differently on density gradient centrifugation. As was expected, ACh was recovered in the fraction containing synaptic vesicles whereas ChAc was confined to fractions containing larger membranes (Fonnum, 1967). The membrane bound enzyme was recovered in a non-occluded form. The enzyme could be solubilized from the membranes by increasing the salt concentration and the pH of the suspending media to more physiological values (pH 7.4, 150 mmoles NaCl) (Fonnum, 1967). The release of ChAc was therefore primarily a function of the pH and ionic strength of the suspending media and the species investigated. About 65% of ChAc was released from membranes by the

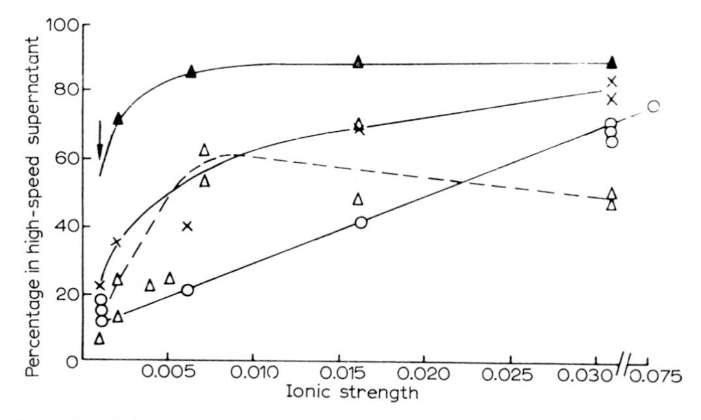

Fig. 1. Release into the high speed supernatant of ChAc from crude synaptosome preparation from different species as a result of suspension in water, followed by adjustment of ionic strength with NaCl. ▲ pigeon, × guinea pig, △ rabbit, ○ rat. (Reproduced from *Biochem. J.* (1967), F. Fonnum, **103**, 262–270.)

addition of NaCl, KCl, $MgCl_2$, $CaCl_2$, ACh, choline, thiols and adenosine phosphates at a concentration corresponding to an ionic strength of 0.01. CoA, acetyl-CoA, Dyflos and eserine did not promote any release of the enzyme in the low concentration tested ($<10^{-4}$ M) (Fonnum, 1968).

A similar mechanism for the release of ChAc was found with different species and with samples from different regions of the brain (Table II).

If ChAc was released from disrupted synaptosomes by adjustment of ionic strength and pH, it could be re-bound to membranes by passing the suspension through Sephadex columns, thereby altering the ionic environments (Fig. 2). Treatment of the re-bound enzyme with Triton X-100 showed that the enzyme was present in a non-

TABLE II

RELEASE OF CHOLINE ACETYLTRANSFERASE AND LACTATE DEHYDROGENASE (LDH) FROM SYNAPTOSOMES ISOLATED FROM DIFFERENT PARTS OF THE RABBIT BRAIN AND FROM DIFFERENT SPECIES AFTER HYPO-OSMOTIC TREATMENT.

		Proportion solubilized (% of total recovered)			
		A		*B*	
Species	*Tissue*	*ChAc*	*LDH*	*ChAc*	*LDH*
Rabbit	Cortex	35	55	65	80
	Caudate nuclei	15	42	60	73
	Hippocampus	22	48	84	78
	Medulla + pons	28	82	57	72
Rat	Cerebra	15	55	70	85
Guinea pig	Cortex	25	64	82	81
Pigeon	Cerebra	72	26	90	59

A crude synaptosome pellet was disrupted by suspension in water (10 ml/g of original tissue); samples were either A, centrifuged immediately or B, adjusted to pH 7.4 and I 0.03 with NaCl, and then centrifuged. Recoveries were 90–100%.

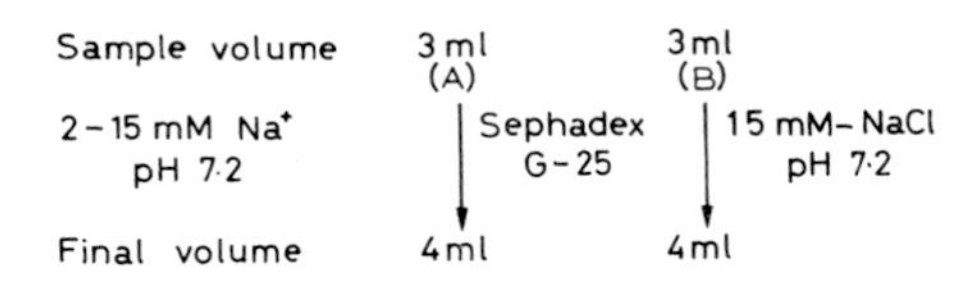

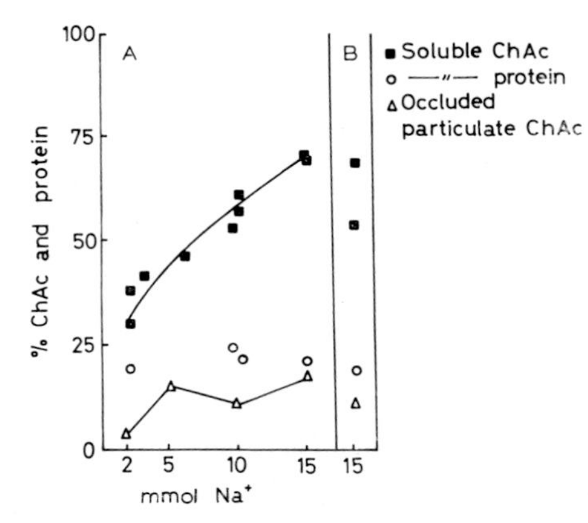

Fig. 2. Binding of ChAc (■) and protein (○) to membranes from synaptosome fraction at various ionic strengths. A, Gelfiltration of membranes and ChAc on Sephadex columns equilibrated with 1 mmole sodium phosphate buffer plus NaCl to give final ionic strength and pH 7.2; and B, Dilution of sample with water to final ionic strength 0.015 and pH 7.2. (Reproduced from *Biochem. J.* (1968), F. Fonnum, **109**, 389–398.)

occluded form. The re-binding could therefore not be accounted for by occlusion of enzyme into ghost particles. The proportion of particulate ChAc obtained after binding experiments was similar to the proportion of particulate ChAc obtained after release experiments provided that the pH and ionic strength were the same in both types of experiments (Fonnum, 1968).

Even more important was the finding that the ChAc re-bound to membranes behaved similarly on density gradient centrifugation to the ChAc obtained directly after hypo-osmotic rupture of synaptosomes (Fig. 3). The binding experiments therefore led to the conclusion that the binding of ChAc to membranes was a reversible process and that the proportion of particulate ChAc was dependent upon the pH, ionic strength and species of the enzyme. A further conclusion was that the binding of ChAc observed after hypo-osmotic treatment could be an artifact due to the lowering of pH and ionic strength accompanying the hypo-osmotic treatment.

Surface charge of ChAc

The surface charge of ChAc was investigated by binding partially purified ChAc

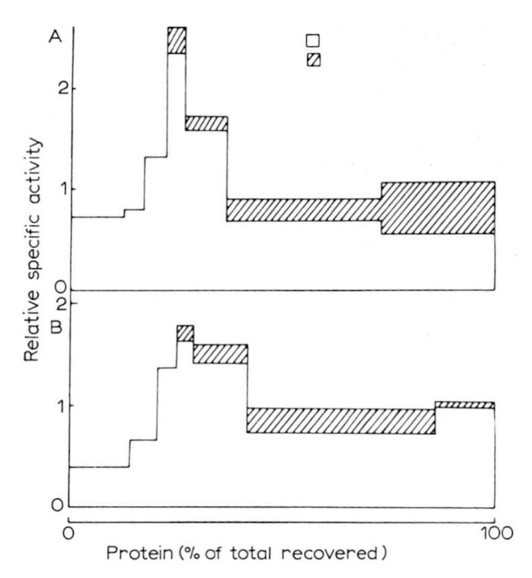

Fig. 3. Distribution of non-occluded (□) and occluded (////) ChAc in fractions separated by discontinuous density gradient centrifuging. The blocks correspond to the fractions 0-I described by Whittaker *et al.* (1964). A, Suspension of hypo-osmotically treated synaptosomes; and B, After binding of soluble ChAc to synaptosome membranes. (Reproduced from *Biochem. J.* (1968), F. Fonnum, **109**, 389–398.)

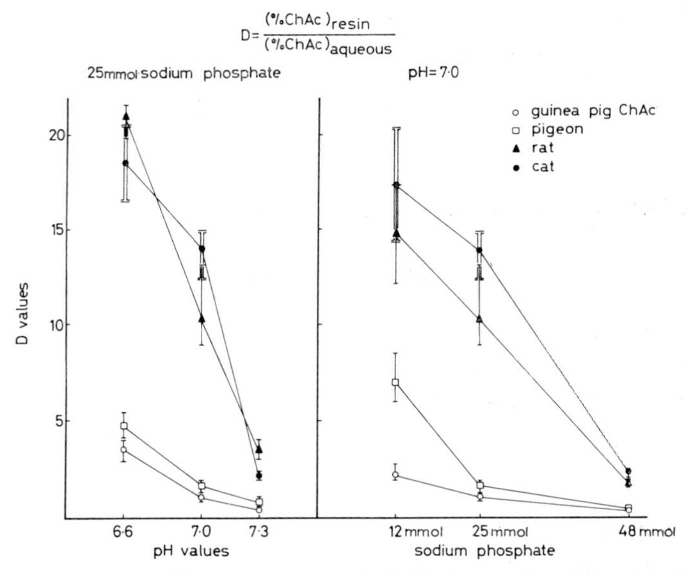

Fig. 4. Binding of partially purified ChAc to CM-Sephadex at varying pH and sodium phosphate buffer concentration.

to CM-Sephadex (C-50) and to Amberlite CG 50-II (Fonnum, 1970). There was no indication that the purification of the enzyme led to any change in enzyme conformation or properties. ChAc was selected from two species (pigeon and guinea-pig) where the enzyme was easily released from membranes, and two species (rat and cat) where the enzyme was strongly bound to membranes.

The binding of ChAc to the ion exchanger was dependent upon pH and ionic strength in the same way as binding of the enzyme to membranes (Fig. 4). When the binding was described in each case in terms of a distribution coefficient D

$$D = \frac{(\%\text{ChAc})\ \text{bound}}{(\%\text{ChAc})\ \text{soluble}}$$

it was found that the D value for the binding of ChAc to CM-Sephadex and Amberlite CG-50 II was 10 times larger than that for the binding of the enzyme to synaptosome membranes at a similar pH and ionic strength. This merely reflects the greater number of binding sites in the cation exchanger experiments. The D value of rat ChAc decreased by a factor of 6 going from pH 6.7 to 7.3 for binding to CM-Sephadex and by a factor of 3–5 for binding to membranes. In both cases the proportion of bound ChAc was independent of ChAc concentration.

More important was the fact that the enzymes from different species could be separated into two groups according to their affinity for the ion exchanger. The enzymes from pigeon and guinea pig brain were less readily absorbed to CM-Sephadex than the enzyme from rat and cat brain (Fig. 4). In agreement, ion exchange chromatography of ChAc on CM-Sephadex showed that the enzyme from pigeon was eluted prior to the enzyme from rat. The enzymes from rat and cat brain therefore had stronger positive surface charges than the enzymes from pigeon and guinea-pig brain.

The experiment provided a molecular basis for explaining the absorption of ChAc to membranes: ChAc was found to be a positively charged molecule and was therefore attracted by negatively charged membranes. Species differences in positive surface charge of enzymes explained the species differences observed in membrane affinity (Fig. 1).

Isoelectric focusing experiments

The molecular heterogeneity of ChAc from different species was further investigated by isoelectric focusing of partially purified enzyme preparations. The isoelectric focusing experiments were run (Malthe–Sørenssen and Fonnum, 1972) with a constant load of 0.5 W for the pH gradient 3–10 and 0.75 W for the pH gradient 6–9. Constant current was usually obtained after 36 h but the column was stopped after 46 h.

The activity of ChAc from pigeon brain was recovered as a single peak with isoelectric point 6.5 $\pm$ 0.1 (Fig. 5a). The peak of enzyme activity did not change on re-focusing in a second gradient nor was the peak changed by the presence of 3 M urea to decrease any glass wall effects or protein–enzyme interactions. ChAc from guinea-pig brain was also recovered as a single peak with a slightly higher isoelectric point, 6.7 $\pm$ 0.1 (Fig. 5b). The position of this peak was also unaltered by the presence of urea or by re-focusing.

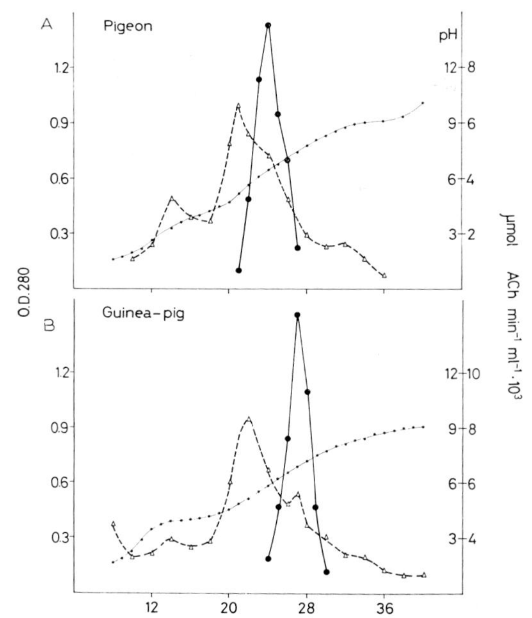

Fig. 5. Isoelectric focusing of A, ChAc from pigeon brain; and B, ChAc from guinea-pig brain in pH gradient 3–10. (●) ChAc activity, (△) protein, (...) pH.

In contrast to the enzymes from pigeon and guinea-pig brains the enzymes from rat and cat brains were distributed over a rather broader pH range. The enzymes from rat brain (Malthe–Sørenssen and Fonnum, 1971a) were distributed as two distinct peaks at pH 7.5 to 7.8 and 8.3 in a pH 3–10 gradient (Fig. 6a). The identity of two separate peaks was established by isolating the most active fractions in the two peaks and re-focusing them in two separate pH 6–9 gradients (Fig. 6b). The results showed that the two new peaks did not overlap and therefore corresponded to two or more different forms of the enzyme. If the original enzyme preparation was run directly in a pH 6–9 gradient, 3 separate peaks of ChAc activity were obtained with isoelectric points 7.5, 7.8 and 8.3 (Fig. 6c).

The necessity of re-focusing experiments was demonstrated with an extract of enzyme from rat caudate nuclei. In this extract a fourth peak was sometimes obtained at pH 6.5 to 7.0. The peak disappeared on re-focusing and the enzyme activity was recovered in the normal range of pH 7.5 to 8.3. This peak is therefore probably an artifact resulting from protein–enzyme interaction.

ChAc from cat brain was separated into two distinct peaks of isoelectric point 7.0 and 8.3 and a third less distinct peak at pH 7.8 (Fig. 7). The identity of the two main peaks was established by re-focusing experiments.

So far we have not obtained any evidence for differences in kinetic behaviour or affinity constants for the different molecular forms of ChAc. The K_m constant for choline was for ChAc from guinea pig, rat isoenzymes and cat 0.75 to 0.85 mmoles, whereas the value was slightly less (0.45 mmoles) for pigeon ChAc. The results are not very surprising since ChAc from such different sources as *Lactobacillus* and the

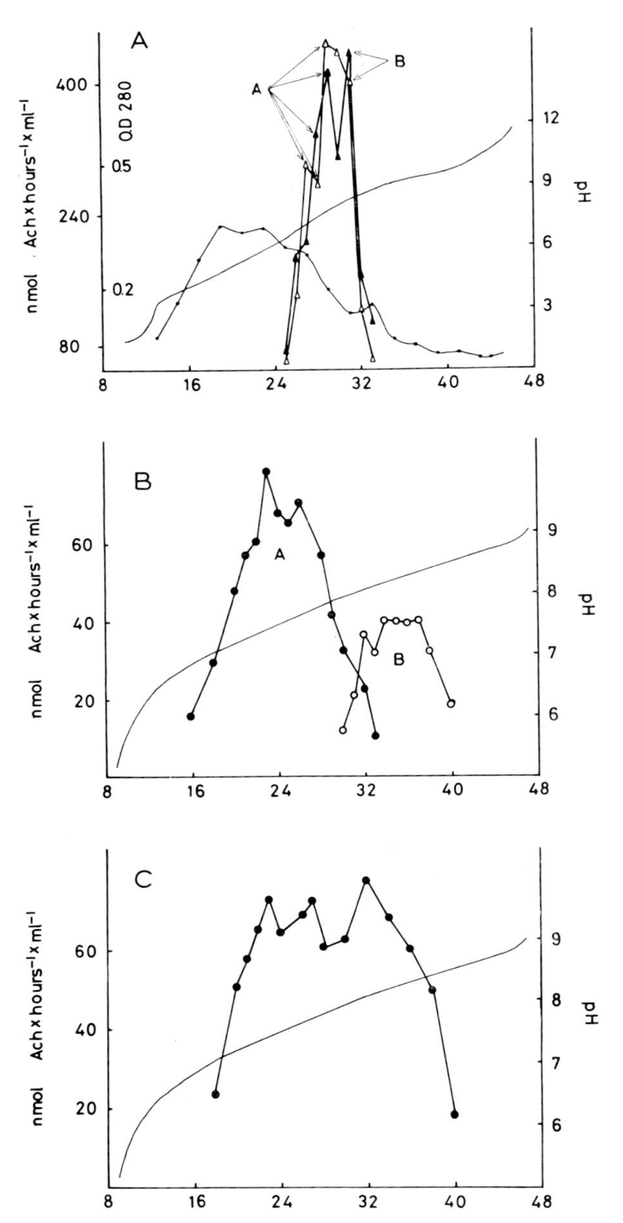

Fig. 6. Isoelectric focusing of ChAc from rat brain. (▲, △, ●, ○) ChAc activity, (—●—) protein, (—) pH. A, Isoelectric focusing in pH 3–10 gradient; B, Re-focusing of the two peaks in pH 6–9 gradient in two separate columns; and C, Isoelectric focusing in pH 6–9 gradient.

calf caudate nucleus showed similar affinity constants for choline and acetyl-CoA (White and Cavallito, 1970). So far a difference in charge is the only criterion which distinguishes the different forms of ChAc.

Isoenzymes differ frequently in their subcellular and cellular localization. Extract

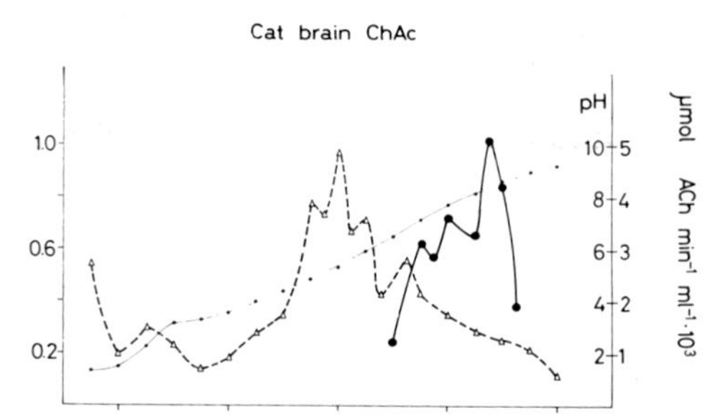

Fig. 7. Isoelectric focusing of ChAc in pH 3–10 gradient from freshly prepared cat brain. (●) ChAc activity, (△) protein, (—) pH.

from the rat hippocampus, caudate nucleus and sciatic nerve contained all 3 ChAc isoenzymes. To investigate if the isoenzymes had different affinity for membranes, synaptosomes were prepared from the rat cortex by density gradient centrifugation (Fonnum, 1968) and then hypo-osmotically treated with 2 mmoles sodium phosphate buffer, pH 7.0. Centrifugation of the disrupted synaptosomes at 100,000 × *g* for 60 min gave a supernatant which contained an easily releasable fraction of ChAc. By further treatment of the pellet with 20 mmoles sodium phosphate buffer, a fraction of ChAc with higher membrane affinity was obtained. Both samples were subjected to isoelectric focusing and the results show (Fonnum and Malthe–Sørenssen, unpublished observations) that the isoenzyme with the isoelectric point 7.5 dominates in the easily released ChAc fraction and the isoenzyme with the isoelectric point 8.3 dominates in the fraction of ChAc with higher membrane affinity (Fig. 8).

Isoelectric focusing experiments therefore confirm the result from the binding of enzyme to the synthetic ion exchanger in that ChAc from rat and cat brain have a higher isoelectric point and therefore stronger positive charge than ChAc from pigeon and guinea-pig brain. The experiments indicate further that in some species closely related molecular forms of ChAc exist and that these forms differ in their membrane affinity.

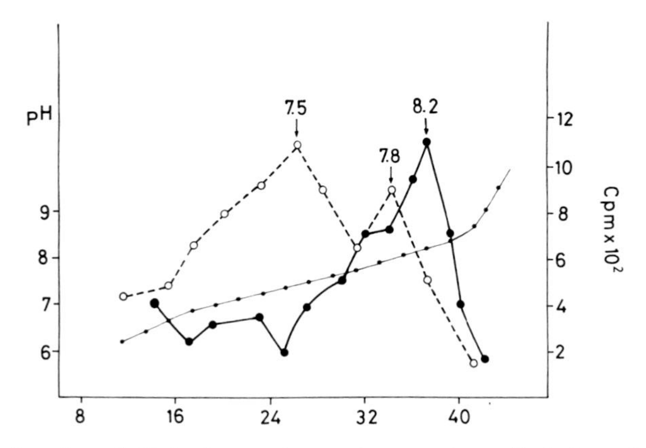

Fig. 8. Isoelectric focusing of ChAc from rat synaptosomes in a pH 6–9 gradient. (—○—) easily releasable ChAc, (—●—) membrane bound ChAc.

Electron microscopy histochemistry of ChAc

The histochemical procedure of Burt (1970) as modified by Kasa *et al.* (1970) has provided another method for investigating the ultrastructural localization of the enzyme. Due to the ease with which the enzyme was released from the tissue at high ionic strength and pH, a precipitate was only formed in the presence of 30 mmoles cacodylate buffer, pH 5.9. Under these conditions 85% of the rat ChAc was bound to membranes. The results (Kasa, 1972) showed that the enzyme was localized all over the cytoplasma of the neurone.

The enzyme was never localized inside any subcellular organelles such as tubules or vesicles but was bound to the outside of different membrane structures including the endoplasmic reticulum, neurotubules, neurofilaments, nerve terminal membranes, vesicles and even mitochondria. This technique therefore confirms the finding from subcellular fractionation that ChAc is soluble under more physiological conditions and tends to be bound to a heterogeneous population of membrane structures at a lower ionic strength and pH.

Different pools of ACh

Homogenization of brain tissue in iso-osmotic sucrose in the presence of a cholinesterase inhibitor gives rise to a soluble pool of ACh that is probably derived from disrupted cholinergic cell bodies, axons and dendrites. Homogenization of peripheral axons (Whittaker *cited* Hebb and Silver, 1963) also gives rise to largely a soluble form of ACh.

When synaptosomes are hypo-osmotically treated, ACh is obtained in a hypo-osmotically labile and hypo-osmotically stable form (Whittaker, 1959). The hypo-osmotically labile form is assumed to be derived from the cytoplasma of the nerve terminals and the hypo-osmotically stable form is assumed to be derived from the vesicles. Different specific labelling of ACh in the 3 pools after infiltration of labelled choline into the cerebral cortex (Chakrin and Whittaker, 1969) constitutes proof that the 3 pools exist *in vivo*.

The binding of ACh to synaptic vesicles has been studied in some detail. ACh was released from the vesicles by a temperature-dependent mechanism (Barker *et al.*, 1967). It was not spontaneously released by the action of cations (Matsuda *et al.*, 1968; Takeno *et al.*, 1969). ACh was still bound to vesicles after gelfiltration in iso-osmotic sucrose, but was lost after gelfiltration in water (Marchbanks, 1968). These studies indicate that the vesicular ACh is protected behind a semipermeable membrane.

Synthesis of ACh from labelled precursors

Injection of labelled choline into the exposed cortex of anaesthetized animals led to a higher specific labelling of the cytoplasmic ACh than of the vesicular ACh. (Chakrin and Whittaker, 1969; Barker *et al.*, 1970). This indicates that the synthesis of ACh occurs in the cytoplasma and therefore that ChAc is localized there.

Incubation of synaptosome preparations with labelled choline and isolation of the different constituents of the synaptosome by hypo-osmotic rupture and density gradient centrifugation showed labelling of the cytoplasmic ACh but very little labelling of the vesicular ACh ($< 2\%$) (Marchbanks, 1969). Incubation of synaptosomes with labelled ACh showed that the cytoplasmic ACh but not the vesicular ACh was readily exchanged with external ACh.

These results are contrary to those of Kaita *et al.* (1970) who found that if synaptosomes were incubated with labelled ACh only 1.5% of ACh was recovered in the vesicle fraction. If, however, synaptosomes were incubated with labelled choline, 15% of the labelled ACh was recovered in the vesicle fraction. The latter authors isolated their vesicles by hypo-osmotic rupture and subsequent differential centrifugation, a procedure that gives less pure fractions. The authors concluded that most of the ACh was synthesized in cytoplasma but that a small proportion (15%) was synthesized in the vesicles. One cannot, however, directly compare results from the uptake of labelled ACh with those from the conversion of labelled choline to labelled ACh. There is at present no evidence that ACh is taken up into cholinergic synaptosomes only, whereas labelled choline is only converted to ACh in cholinergic synaptosomes. Since only cholinergic vesicles may possess the ability to take up ACh, and assuming that only 10% of the synaptosomes in the cortex are cholinergic, one would expect that an uptake of 1.5% of ACh into all vesicles would correspond to an uptake of 15% of ACh into cholinergic vesicles. The experiments of Kaita *et al.* (1970) therefore do not exclude the idea that ACh may be synthesized in the cytoplasma and then taken up into the vesicles.

Similar objections may be raised to experiments where labelled ACh was allowed to diffuse into the cerebral cortex and ACh was collected afterwards by stimulation. Since ACh would be taken up by both cholinergic and noncholinergic nerve terminals and probably also would be spontaneously released from such structures, it would be very difficult to detect any increase of released labelled ACh on stimulation.

Synthesis and storage of vesicular ACh

Subcellular fractionation of ChAc, studies of the electric charge of ChAc and electron microscopy histochemistry all favour the view that ChAc is distributed throughout the cytoplasm. The enzyme has a high affinity for membranes but under physiological conditions with regard to pH and ionic strength the soluble form of the enzyme is expected to dominate.

If ACh is synthesized extravesicularly as all experiments seem to indicate at present, there must exist a mechanism for the uptake of ACh into the synaptic vesicles. This mechanism has not yet been convincingly established. Experiments *in vivo* with infusion of labelled choline give labelling of ACh in synaptic vesicles (Chakrin and Whittaker, 1969; Barker *et al.*, 1970) whereas the results from isolated synaptosome preparations are controversal.

The only convincing evidence for an uptake of ACh into vesicles *in vitro* has been provided by Guth (1969) who demonstrated that vesicles took up ACh within seconds. This mechanism was very sensitive and decreased rapidly with time, an explanation of why other investigators have failed to demonstrate it. However, there is still no evidence that this pool of ACh is present behind a semipermeable membrane and is not simply a result of diffusion or ion exchange binding. Alternative explanations for the formation of vesicular ACh can be suggested by taking into account the membrane affinity of ChAc. A physiological function of this membrane affinity has become more probable because the isoenzymes of rat differ in their membrane affinities. If a small part of ChAc is linked to the outside of the synaptic vesicles, the ACh synthesized by it will be in an excellent position for being taken up into the vesicles. Alternative explanations are that part of ChAc are bound to tubules or neurofilaments and that these structures are connected with the loading of transmitters.

Several investigators (Collier, 1969; Whittaker, 1970) have claimed that the newly synthesized ACh is preferentially released from the autonomic ganglion and brain cortex. Such results may be expected if a small part of ChAc is bound to the outer synaptosome membrane near the point of entry of choline. It can then utilize this choline, before it mixes with the endogeneous choline, for synthesis of ACh.

Since ChAc is relatively easily released from synaptosome membranes, it is necessary to keep in mind that only a minor portion of the enzyme can be bound to membranes. If there are microenvironmental changes in pH within the terminal (*e.g.*, due to the hydrolysis of ACh) a larger proportion of the enzyme could be membrane bound.

During stimulation of cholinergic nerves (*e.g.*, phrenic nerve diaphragm preparation) there is a 2–5 fold increase in released ACh (Krnjević and Mitchell, 1961; Mitchell and Silver, 1963; Schmidt, Szilagyi, Alkon and Green, 1970, Szerb, 1971).

The spontaneous release of ACh from rat diaphragm has been studied at various temperatures and by soaking the tissue in Ringer's solution containing a high K^+ concentration (Mitchell and Silver, 1963). The results suggest that only a minor part of the spontaneously released ACh was associated with miniature end plate potentials. The main proportion of spontaneously released ACh could therefore come from the large pool of cytoplasmic ACh and could simply be an overflow mechanism for regulating the level of ACh in the terminals. During stimulation this flow of ACh may be directed towards the vesicle.

CONCLUSION

Choline acetyltransferase can be obtained in 4 different forms on subcellular fractionation of brain tissue. If the tissue is homogenized in sucrose the enzyme is obtained either in an occluded particulate form (the synaptosomes) or in the soluble form (disrupted cell bodies, axons and dendrites). The ratio of soluble to particulate ChAc in sucrose homogenate varies for the different regions of the brain and reflects regional differences in distribution of cholinergic structures. If the synaptosomes are hypo-osmotically treated, the enzyme is obtained either in a soluble (cytoplasma) or a particulate non-occluded (membrane bound) form. The two latter forms are reversibly interchangeable depending upon the pH and ionic strength of the suspending media. The different membrane affinities of ChAc from different species are dependent upon the net surface charge of the enzyme. It is demonstrated that ChAc from cat and rat have a higher positive surface charge than those from pigeon and guinea pig. In agreement, isoelectric focusing shows that the enzymes from rat and cat brains have higher isoelectric points than those from pigeon and guinea-pig brains. In addition, ChAc from rat and cat brains consists of two or three different isoenzymes with different isoelectric points and different membrane affinities. Different possible physiological functions of the membrane affinities for ChAc are discussed.

SUMMARY

The compartmentation of choline acetyltransferase (ChAc) and its consequences for the synthesis storage and release of acetylcholine (ACh) within the neurone are discussed. The present experiments showed that the binding of ChAc to membranes was a reversible process primarily dependent upon the pH, ionic strength and the ionic properties of ChAc. The binding of ChAc to membranes resembled the ionic attraction between ChAc and a cationic exchange resin in all aspects. ChAc from different species had a different surface charge and a different isoelectric point. Isoelectric focusing of ChAc showed that only one form of ChAc was present in pigeon and guinea pig, whereas in rat and cat there were three and two (respectively) different molecular forms. Different molecular forms of ChAc had different membrane affinity. The literature on the electron microscopy histochemistry of ChAc is reviewed.

REFERENCES

BARKER, L. A., AMARO, J. AND GUTH, P. S. (1967) Release of acetylcholine from isolated synaptic vesicles. 1. Methods for determining the amount released. *Biochem. Pharmacol.*, **16**, 2181–2189.

BARKER, L. A., DOWDALL, M., ESSMAN, W. B. AND WHITTAKER, V. P. (1970) In *Drugs and Cholinergic Mechanisms in the CNS*, E. HEILBRONN AND A. WINTER (Eds.), Res. Inst. of Natl. Def., Stockholm, Almqvist and Wiksell, Stockholm, pp. A3–214.

BURT, A. M. (1970) A histochemical procedure for the localization of choline acetyltransferase activity. *J. Histochem. Cytochem.*, **18**, 408–415.

CHAKRIN, L. W. AND WHITTAKER, V. P. (1969) The subcellular distribution of (N-ME-^{3}H) Acetylcholine synthesized by brain *in vivo*. *Biochem. J.*, **113**, 97–107.

COLLIER, B. (1969) The preferential release of newly synthesized transmitter by a sympathetic ganglion. *J. Physiol. (Lond.)*, **205**, 341–352.

DE ROBERTIS, E. (1967) Ultrastructure and cytochemistry of the synaptic region. *Science*, **156**, 907–914.

DE ROBERTIS, E., RODRIGUEZ DE LORES ARNAIZ, G., SALGANICOFF, L., PELLEGRINO DE IRALDI, A. AND ZIEHER, L. M. (1963) Isolation of synaptic vesicles and structural organization of the acetylcholine system within brain nerve endings. *J. Neurochem.*, **10**, 225–235.

FONNUM, F. (1966a) A radiochemical method for the estimation of choline acetyltransferase. *Biochem. J.*, **100**, 479–484.

FONNUM, F. (1966b) Is choline acetyltransferase present in synaptic vesicles? *Biochem. Pharmacol.*, **15**, 1641–1643.

FONNUM, F. (1967) The "compartmentation" of choline acetyltransferase within the synaptosome. *Biochem. J.*, **103**, 262–270.
FONNUM, F. (1968) Choline acetyltransferase, binding to and release from membranes. *Biochem. J.*, **109**, 389–398.
FONNUM, F. (1970) Surface charge of choline acetyltransferase from different species. *J. Neurochem.*, **17**, 1095–1100.
GUTH, P. S. (1969) Acetylcholine binding by isolated synaptic vesicles *in vitro*. *Nature (Lond.)*, **224**, 384–385.
HEBB, C. O. AND SILVER, A. (1963) The effect of transection on the level of choline acetylase in the goat's sciatic nerve. *J. Physiol. (Lond.)*, **169**, 41*P*–42*P*.
HEBB, C. O. AND SMALLMAN, B. N. (1956) Intracellular distribution of choline acetylase. *J. Physiol. (Lond.)*, **134**, 385–92.
HEBB, C. O. AND WHITTAKER, V. P. (1958) Intracellular distributions of acetylcholine and choline acetylase. *J. Physiol. (Lond.)*, **142**, 187–96.
KAITA RITCHIE, A. AND GOLDBERG, A. M. (1970) Vesicular and synaptoplasmic synthesis of acetylcholine. *Science*, **169**, 489–490.
KASA, P., MANN, S. P. AND HEBB, C. (1970) Localization of choline acetyltransferase. Ultrastructural localization in spinal neuron. *Nature (Lond.)*, **226**, 814–816.
KÁSA, P. (1972) Ultrastructural localisation of choline acetyltransferase and acetylcholinesterase in central and peripheral nervous tissue. *Progr. Brain Res.*, **34**, in press.
KRNJEVIĆ, K. AND MITCHELL, J. F. (1961) The release of acetylcholine in the isolated rat diaphragm. *J. Physiol. (Lond.)*, **155**, 246–262.
MALTHE–SØRENSEN, D. AND FONNUM, F. (1971) Multiple forms of choline acetyltransferase from rat brain. *Nature (Lond.)*, **229**, 127.
MALTHE–SØRENSEN, D. AND FONNUM, F. (1972) Multiple forms of ChAc in several species demonstrated by isoelectric focusing. *Biochem. J.*, in press.
MARCHBANKS, R. M. (1968) Exchangeability of radioactive acetylcholine with the bound acetylcholine of synaptosomes and synaptic vesicles. *Biochem. J.*, **106**, 87–95.
MARCHBANKS, R. M. (1969) The conversion of ^{14}C-choline to ^{14}C-acetylcholine in synaptosomes *in vitro*. *Biochem. Pharmacol.*, **18**, 1763–1766.
MATSUDA, T., HATA, F. AND YOSHIDA, H. (1968) Stimulatory effect of Na^+ and ATP on the release of acetylcholine from synaptic vesicles. *Biochim. biophys. Acta (Amst.)*, **150**, 739–741.
MCCAMAN, R. E., RODRIGUEZ DE LORES ARNAIZ, G. AND DE ROBERTIS, E. (1965) Species differences in subcellular distribution of choline acetylase in the CNS. A study of choline acetylase, acetylcholinesterase, 5-hydroxytryptophan decarboxylase, and monoamine oxidase in four species. *J. Neurochem.*, **12**, 927–935.
MITCHELL, J. F. AND SILVER, A. (1963) The spontaneous release of acetylcholine from the denervated hemidiaphragm of the rat. *J. Physiol. (Lond.)*, **165**, 117–129.
SCHMIDT, D. E., SZILAGYI, P. I. A., ALKON, D. L. AND GREEN, J. P. (1970) A method for measuring nanogram quantities of acetylcholine by pyrolysis–gas chromatography: The demonstration of acetylcholine in effluents from the rat phrenic nerve–diaphragm preparation. *J. Pharmacol. exp. Ther.*, **174**, 337–345.
SILVER, A. (1967) Cholinesterase of the central nervous system with special reference to the cerebellum. *Int. Rev. Neurobiol.*, **10**, 57–109.
SZERB, J. C. (1971) The effect of atropine on the metabolism of acetylcholine in the cerebral cortex. In *Prog. Brain Res.*, *Vol. 36*, P. B. BRADLEY AND R. W. BRIMBLECOMBE (Eds.), Elsevier, Amsterdam, pp. 159–165.
TAKENO, K., NISHIO, A. AND YANAGIYA (1969) Bound acetylcholine in the nerve ending particles. *J. Neurochem.*, **16**, 47–52.
TEČEK, S. (1966a) On subcellular localization and binding of choline acetyltransferase in the cholinergic nerve endings of that brain. *J. Neurochem.*, **13**, 1317–1327.
TUČEK, S. (1966b) On the question of the localization of choline acetyltransferase in synaptic vesicles. *J. Neurochem.*, **13**, 1329–1332.
WHITE, H. L. AND CAVALLITO, C. J. (1970) Inhibition of bacterial and mammalian choline acetyltransferases by styrylpyrimidine analogues. *J. Neurochem.*, **17**, 1579–1589.
WHITTAKER, V. P. (1959) The isolation and characterization of acetylcholine containing particles from brain. *Biochem. J.*, **72**, 694–706.
WHITTAKER, V. P. (1965) The application of subcellular fractionation techniques to the study of brain function. *Progr. Biophys. molec. Biol.*, **15**, 39–96. Universitetsforlaget, Oslo.

WHITTAKER, V. P. (1970) The vesicle hypothesis. In *Excitatory Synaptic Mechanisms*, P. ANDERSEN AND J. K. S. JANSEN (Eds.), pp. 67–76.
WHITTAKER, V. P., MICHAELSON, I. A. AND KIRKLAND, R. J. A. (1964) The separation of synaptic vesicles from nerve-ending particles (synaptosomes). *Biochem. J.*, **90**, 293–303.

DISCUSSION

SZERB: (1) What is the evidence that the caudate nucleus of the cat contains mostly cholinergic terminals and not any cell bodies? (2) What is the role of intrasynaptosomic cholinesterase in the scheme of ACh synthesis which you propose?

FONNUM: The evidence that cholinergic terminals and not cell bodies are present in the caudate nucleus is based on the electron microscopy of AChE by Lewis and Shute (1967) in that region. Acetylcholinesterase seems to be bound within the terminals and not on the cells, that is within the cell reticular system. Further evidence is given by the experiment of Lewis and Shute which shows that if you cut the pathway between the substantia nigra and the caudate nucleus, the AChE activity seems to disappear.

The second question depends on whether AChE is present on the inside or the outside of the nerve terminals. Most people working with subcellular fractionation have obtained evidence to show that the main part of the AChE must be on the outside (Fonnum, Rodriguez de Lores Arnaiz and Marchbanks, independent studies, unpublished observations). This is mainly based on two kinds of study. (1) If the cholinesterase activity of isolated synaptosomes is measured in sucrose, in water (hypotonic disruption) or by treatment with a detergent such as Triton X-100, the enzyme activity remains unaltered, *i.e.*, the enzyme, unlike ChAc, is not protected behind the membrane. (2) If the synaptosome is loaded with labelled ACh, the latter is not hydrolysed; but if the synaptosomes are broken, ACh diffuses out and is immediately hydrolysed by the AChE. If there is any AChE on the inside of the nerve terminal it must therefore be only a very small amount.

ANSELL: Since ChAc tends to become adsorbed on various membrane fractions during subcellular fractionation procedures, how can one be sure that it is a suitable marker enzyme for synaptosomes?

FONNUM: This is in fact very easily done and we can use the same procedure as for lactic dehydrogenase. If the synaptosome fraction is assayed in the absence of a detergent, about 10–15% of the activity is obtained. The amount increases on treatment with a detergent, thus occluded ChAc can be measured and used as a marker for cholinergic synaptosomes.

BRADLEY: Is there not another possibility, apart from the ones you considered, regarding the presence of ACh in two pools, one soluble and the other in the vesicles? Namely, that the vesicles are artifacts.

FONNUM: That has been suggested quite a number of times and we always bear it in mind, but so far all studies of transmitter substances, that is on biogenic amines and on ACh, seem to show that we have two parts of ACh or of transmitter, one of which is more bound than the other, and it is very easy to think of these as being in the vesicles and in the cytoplasm. But it has never been really proved beyond doubt and one is always a bit sceptical of it. But nothing so far convinces me that this is untrue and I shall go on, as all the textbooks, believing that it is true.

REFERENCES

LEWIS, P. R. AND SHUTE, C. C. D. (1967) The cholinergic limbic system: projections to hippocampal formation, medial cortex, nuclei of the ascending cholinergic reticular system and the subfornical organ and supra-optic crest. *Brain*, **90**, 521–540.

Action of Phospholipase A on Synaptic Vesicles. A Model for Transmitter Release?

EDITH HEILBRONN

Research Institute of Swedish National Defence, Department 1, 172 04 Sundbyberg 4 (Sweden)

Phospholipids are part of the structure of membranes found at cholinergic synapses and they may also be actively involved in chemical transmission. It was shown with $^{32}PO_4^{3-}$ as a precursor that phospholipids of nervous tissue, preferentially phosphatidic acid and phosphatidylinositol but also other phospholipids, particularly phosphatidylethanolamine and to some extent phosphatidylcholine, increase their turnover rate upon electrical stimulation. Only phosphatidylinositol was affected in excised rat sympathetic ganglia and the effect was probably postsynaptic (Larrabee *et al.*, 1963); ganglia *in situ* were similarly affected (Larrabee, 1968). Potassium ion stimulation (Heilbronn and Widlund, 1970), the addition of acetylcholine (ACh) (Hokin and Hokin, 1955, 1958; Durell and Sodd, 1964, 1966; Heilbronn and Widlund, 1970) or ACh-releasing substances, such as atropine and certain psychotomimetic glycollates (Abood and Biel, 1962; Heilbronn and Widlund, 1970), also affect phospholipid turnover in nervous tissue. Phospholipids in brain cortex slices of the rat, turning over more rapidly in the presence of glycollates or added ACh, are located mainly in subcellular fractions that have been heavily enriched in nerve ending particles (Heilbronn and Widlund, 1970). These observations may not all have the same cause but they warrant a more careful examination of the role of phospholipids in synaptic transmission.

Morphological analysis has shown that a phospholipase A from the venom of *Naja naja siamensis* (isolated by Eaker *et al.*, Institute of Biochemistry, University of Uppsala, Sweden) was able to break and digest the external membrane of nerve ending particles, leaving the junction areas seemingly intact (Heilbronn and Cedergren, 1970; Cedergren *et al.*, 1970). It was observed that synaptic vesicles form clusters in the presence of phospholipase A and seem to be reduced in number. This reduction was thought to be either the result of vesicles disappearing through the broken external nerve ending membranes or that of phospholipase A attacking the vesicles themselves. Certainly the formation of clusters indicated a change in the structure of the vesicle membranes. If vesicles were indeed attacked by the enzyme they might release their content of ACh as a consequence. A model for transmitter release could be designed, requiring the presence, at the synapse, of an enzyme able to change the permeability of the vesicle membrane. Experiments were designed to investigate this model.

References pp. 38–39

Isolation of vesicles

The experiments were started with a preparation of cholinergic vesicles, which was obtained from the purely cholinergically innervated electric organ of *Torpedo nobiliana*. The vesicles were collected by zonal centrifugation in a B29 rotor during a stay at the Marine Biological Laboratory, Woods Hole, Mass., U.S.A.* The gradient used was a modification of the sucrose–NaCl gradient described by Barker *et al.* (1970). The fractions containing cholinergic vesicles were identified by assaying acid-boiled samples of each fraction for ACh on the dorsal muscle of the leech. Fractions were also analysed for protein (Lowry *et al.*, 1951), acetylcholinesterase (AChE) (Ellman

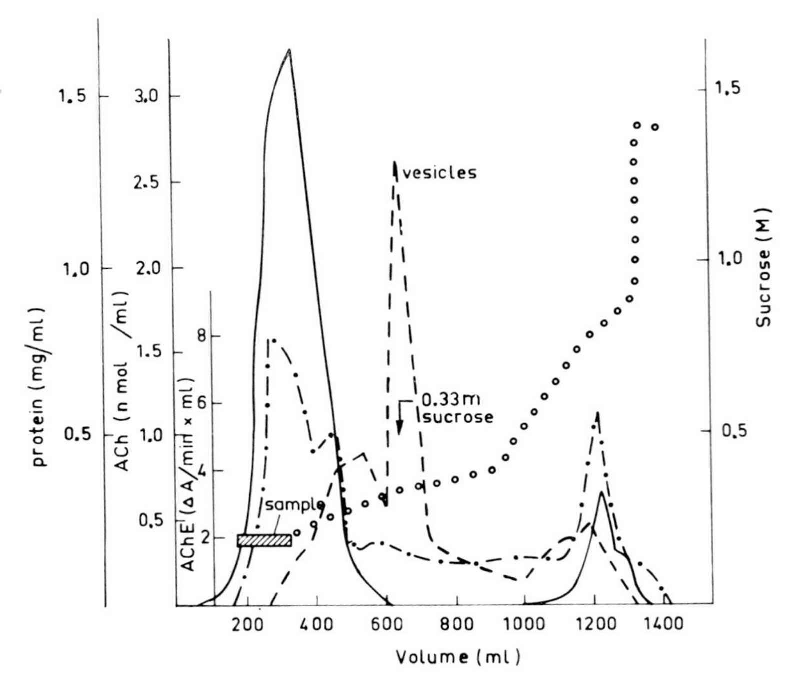

Fig. 1. Isolation of cholinergic vesicles from *Torpedo nobiliana* by zonal centrifugation in a sucrose–NaCl gradient. The large protein peak corresponds to soluble protein, the small one to membrane bound protein. A few particles corresponding to nerve ending particles may have been formed, as indicated by the presence of both AChE and ACh in the small protein fraction. (— — — —) ACh, (—●—●—) AChE, (————) protein, (○○○○○) sucrose gradient.

et al., 1961) and further checked by morphological analysis. The vesicles had a density corresponding to 0.32–0.34 M sucrose. The result of a zonal run is seen in Fig. 1.

Further experiments, performed in Sweden, were carried out with vesicles prepared from the brain cortex of the rat. The tissue was homogenized in 10 times its weight

* The vesicles were collected in collaboration with Drs. S. Hause, Department of Biochemistry, Illinois Inst. of Technology, Chicago, U.S.A., C. A. Price, Department of Agriculture, Rutgers University, New Brunswick, U.S.A. and V. P. Whittaker, Department of Biochemistry, University of Cambridge, Cambridge, England.

of 0.32 M sucrose in the presence of 2×10^{-5} M eserine and 10^{-6} M $CaCl_2$. Nerve ending particles were then isolated according to Whittaker (1959). The 0.9 M sucrose fraction, which contained the nerve ending particles, was diluted to 0.32 M sucrose with deionized water containing eserine and $CaCl_2$ and spun in a Spinco 50 ultracentrifuge at 40,000 rpm for 20 min. The pellet was suspended in 10 vol. of an eserine–$CaCl_2$ solution, homogenized for 2 min at 840 rpm and left in the cold for another 30 min. It was then centrifuged at 9,500 rpm in a Servall centrifuge SS 34 for 20 min. The supernatant was placed on a gradient and spun for 2 h at 25,000 rpm in an SW 40 rotor. The vesicle-containing fraction was used for further experiments with phospholipase A. Fractions mainly containing membrane fragments were saved and later analysed for intrinsic phospholipase A activity.

Effect of phospholipase A on cholinergic vesicles

Suspensions of isolated vesicles were tested for ACh activity before and after acid boiling, the latter method releasing the vesicle-bound ACh. Usually about 50% of the total amount of ACh was found in the suspension medium before the boiling. Freshly prepared vesicles were incubated at room temperature with either 5, 10 or 20 μl of phospholipase A for 30 sec or longer. This resulted in a gradual weakening of the vesicle membranes until the membranes were finally broken down (Figs. 2, 3 and 4). Morphological analysis indicated the formation of distinct pieces of membrane, about one-third the size of the total vesicle membrane. Measurements of the cut vesicle structures seen on the electron micrographs ($\times$ 80,000) give the value of 10.8 ± 1.14 mm for the vesicle membranes and that of 3.4 ± 1.44 mm for the pieces formed upon treatment with phospholipase A. Bioassays of the solutions showed increasing amounts of ACh in the suspension medium in the presence of phospholipase A (Fig. 5). By incubation of the solutions with 1 mg/ml of bovine erythrocyte AChE for 5 or 12 min at room temperature and by gas chromatography and mass spectrometric analysis* (methods according to Hanin and Jenden, 1969; *cf.* Heilbronn *et al.*, 1971), it was established that the observed contractions of the leech dorsal muscle were indeed caused by ACh. Controls treated as described above but without added phospholipase A showed no breakdown of vesicle membranes and did not release ACh into the suspension medium during the course of the experiments (Figs. 2 and 5).

In some experiments vesicle membranes treated with phospholipase A and control vesicles were separated from the ACh in the suspension medium by gelfiltration on a Sephadex G-50 column. Vesicles or enzyme-treated vesicles were identified by protein analysis and came out with the void volume. After acid boiling it was found by bioassay that the enzyme-treated vesicles contained less ACh than the controls.

Assay of nerve ending particle membranes for phospholipase A activity

Preparation of [^{3}*H*]*-lecithin.* Rat brain was incubated with [^{3}H]-choline. After

* Ing G. Lundgren, Department of Toxicology, Karolinska Institutet, Stockholm, Sweden, kindly performed this analysis.

References pp. 38–39

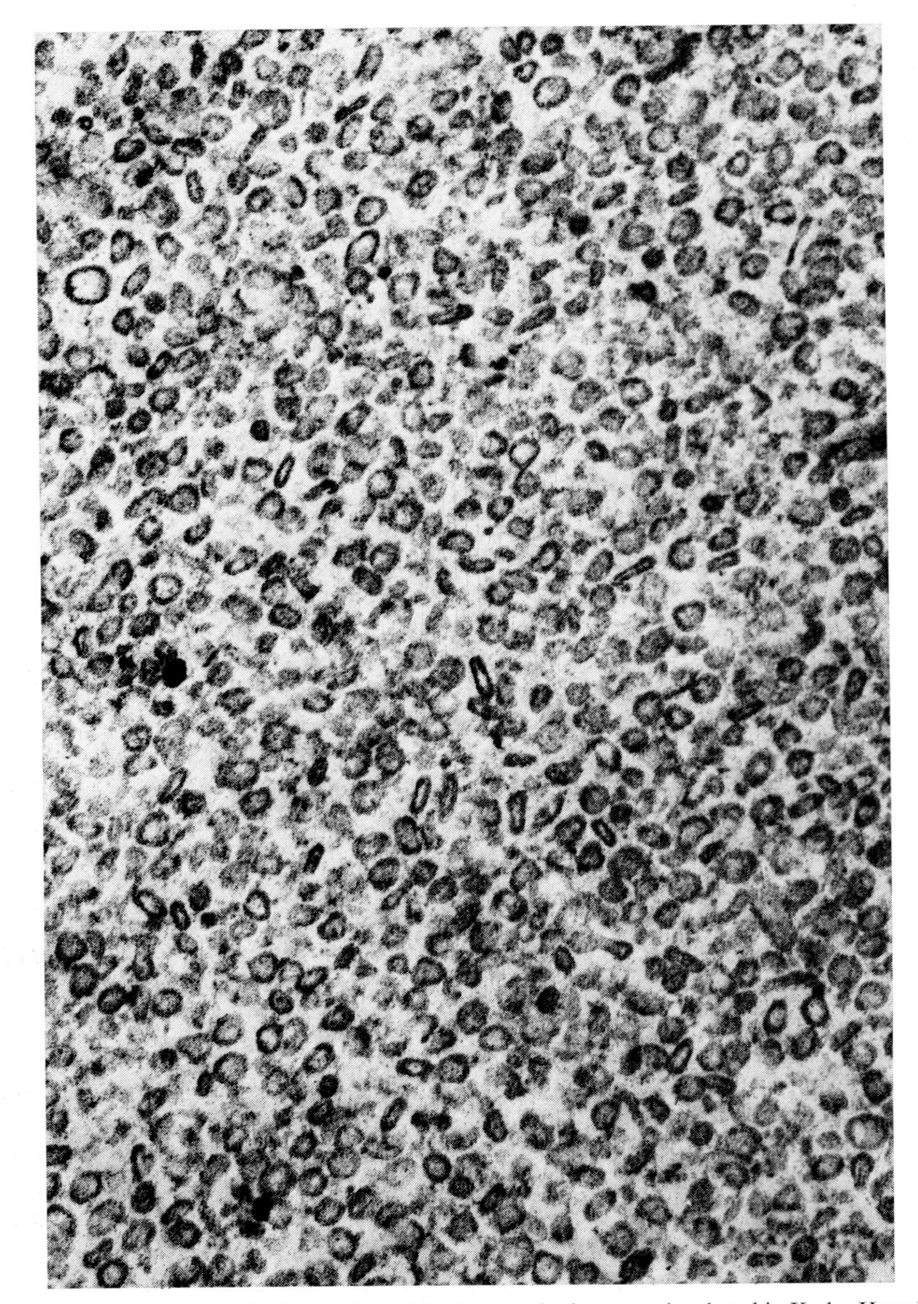

Fig. 2. Electron micrograph of synaptic vesicles from rat brain cortex, incubated in Krebs–Henseleit buffer for 5 min. × 80,000.

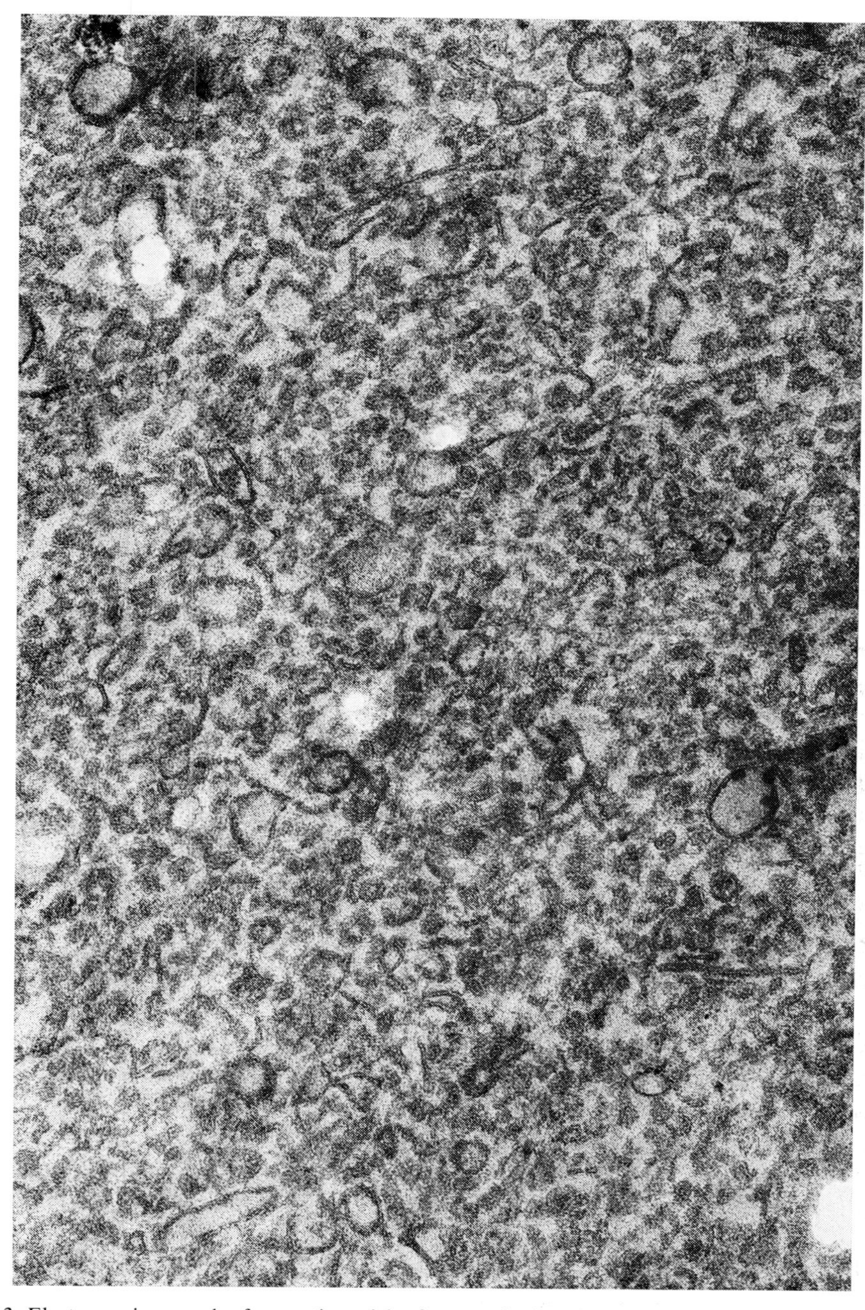

Fig. 3. Electron micrograph of synaptic vesicles from rat brain cortex, incubated in Krebs–Henseleit buffer in the presence of phospholipase A for 30 sec. × 80,000.

References pp. 38–39

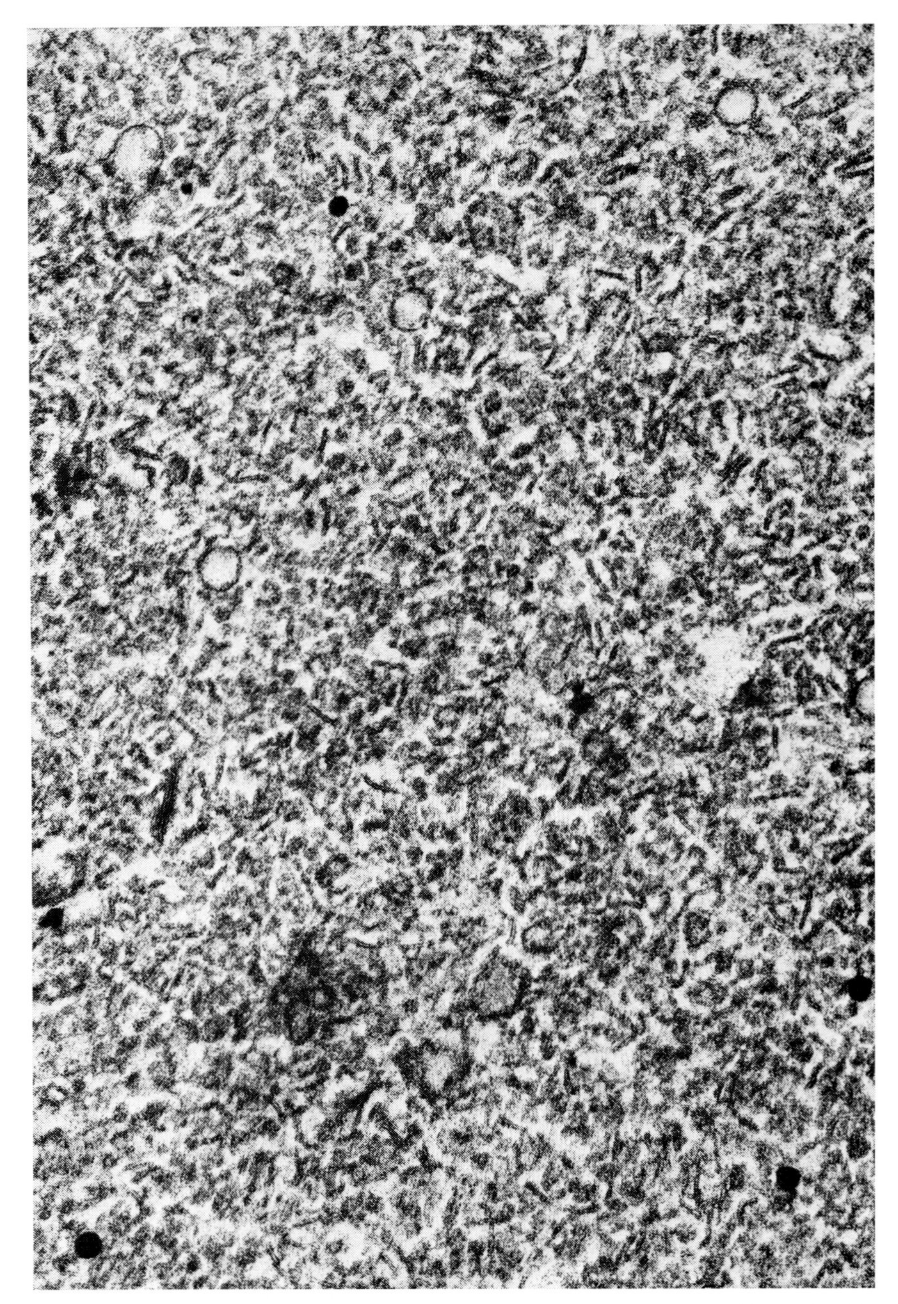

Fig. 4. Electron micrograph of synaptic vesicles from rat brain cortex, incubated in Krebs–Henseleit buffer in the presence of phospholipase A for 5 min. × 80,000.

homogenization, phospholipids were extracted from the tissue as described by Heilbronn and Widlund (1970). After further purification by preparative thin layer chromatography on silica gel H, with chloroform–methanol–acetic acid–water (25:15:4:2 v/v) as the solvent and several extraction steps, the material obtained was dissolved in chloroform. Co-chromatography with authentic lecithin showed only one spot. The specific activity of the [^{3}H]-lecithin obtained was 0.743 mCi/mmole, with a counting efficiency of 28%. *Naja Naja siamensis* phospholipase A was shown to form labelled lysolecithin from this compound.

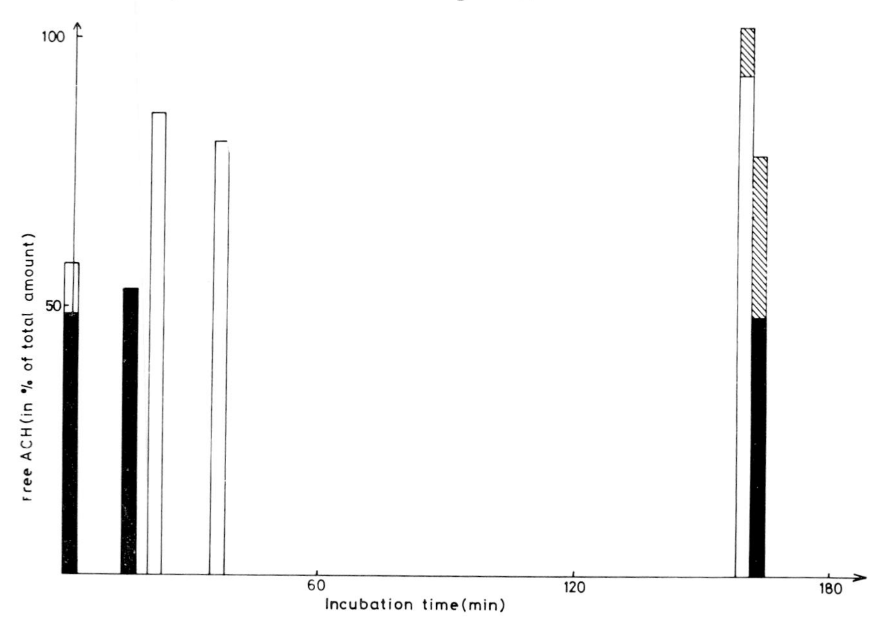

Fig. 5. Biologically active ("free") ACh in a suspension of synaptic vesicles from rat brain cortex as a function of time. (■) control, before acid boiling, (□) vesicles treated with phospholipase A, (////) ACh released from control and phospholipase A-treated vesicles by acid boiling.

Test for phospholipase A activity. Acetone powders of membrane fractions were incubated with [^{3}H]-lecithin at 37°C. The incubation mixture was extracted with methanol–chloroform, the extracted solution was washed and concentrated and the labelled phospholipids were separated and identified by thin layer chromatography in the presence of authentic reference compounds. The spots were scraped off and their radioactivity was measured. It was found that the membrane powders had formed some [^{3}H]-lysolecithin.

Conclusions and discussion

The experiments described show that phospholipase A, isolated from the venom of *Naja Naja siamensis* and able to form lysolecithin from biosynthetically formed labelled lecithin, gradually weakens and breaks down synaptic vesicles from rat brain cortex and cholinergic vesicles from the electric organ of *Torpedo nobiliana*, thereby

References pp. 38–39

releasing ACh into the vesicle suspension. Morphological analysis shows the rapid onset of changes in the structure of the vesicle membranes, followed by their total breakdown. The electron micrographs also indicate the formation of pieces of membranes (seemingly double-lined), about one-third the circumference of whole vesicle membranes in length. This observation may indicate discrete areas in the vesicle membrane which are sensitive either to phospholipase A or to the phospholipid produced by its action. The possible reformation of membranes should of course not be overlooked. The membrane pieces seem however to have lost their capacity to form vesicles. The interference of phospholipase A in repeated leech muscle tests has not yet allowed us to correlate the amounts of ACh released with the different stages of vesicle membrane changes, but the experiments performed suggest that weakened but still vesicle-shaped membranes hold part of the ACh.

From work with beef-heart mitochondrial cristae and *Naja Naja siamensis* phospholipase A it is known that phosphatides are digested in the order phosphatidylethanolamine, phosphatidylcholine and diphosphatidylglycerol (cardiolipin). The majority of the lysophosphatides seems to stay in the membrane. Low level digestion with phospholipase A, sufficient to cause almost complete breakdown of phosphatidylethanolamine and phosphatidylcholine does not produce noticeable changes in membrane structure. Another *Naja Naja* phospholipase A preparation was also shown to attack phosphatidylethanolamine faster than phosphatidylcholine (Gallai–Hatchard and Gray, 1968)

Synaptic vesicle fractions are known to contain phospholipids such as lecithin, sphingomyelin and possibly some lysolecithin (Burton and Gibbons, 1964). A complete analysis of the phospholipid content of cholinergic vesicles is not yet available. The changes obtained in the phospholipid composition of the synaptic vesicles after treatment with phospholipase A are under investigation, as is the extent to which they may be due to enzymatically formed lysophosphatides. Protein that may have been released by a sequential enzymic digestion of the vesicles have not yet been analysed. Clearly, however, the fatty acid parts of the membrane phospholipids have to be intact and in place if the membrane structure is to be retained, as hydrophobic forces to a large extent seem to be holding membranes together. This may be the case at the beginning of phospholipase A action. It was earlier found that phospholipase C, which splits off phosphorylated bases from phospholipids and leaves diglycerides, is much less effective in decreasing the ACh content of brain cortex slices than phospholipase A (Heilbronn, 1969), an observation that suggests that the diglycerides formed to a large extent could maintain the membrane structures of the nerve endings. This fits in with observations of Lenard and Singer (1968) on the red cell membrane and those of Condrea and Rosenberg (1968) and Rosenberg and Condrea (1968) on the squid giant axon.

From studies on the incorporation of [^{3}H]-leucine into proteins of the synaptic vesicles and nerve ending membranes, a half life of 20–22 days was calculated (Von Hungen *et al.*, 1968). Lapetina *et al.* (1969) found a more rapid turnover for brain phospholipids, though different parts of the molecule seemed to turn over at different rates. Acylation of lysophospholipids to diacyl-phospholipids seems to be carried out by brain tissue (Webster, 1965). The evidence summarized in this article indicates that synaptic vesicles are substrates for a phospholipase A obtained from *Naja Naja siamensis*. Short incubation with low amounts of the enzyme changes the structure of the still intact membranes, probably due to a beginning selective breakdown of membrane phospholipids. The specificity of the enzyme points to phosphatidylethanolamine as the first compound to be attacked. Further treatment of vesicular membranes results in the breakdown of the vesicular structure and the formation of small open membrane pieces is observed, indicating spots in the original membrane particularly sensitive to the enzyme or the lysophosphatides formed by its action. A release of vesicle-bound ACh was observed as a consequence of the enzyme treatment. Phospholipases are often Ca^{2+}-dependent and so is the release of ACh. As subcellular fractions obtained from rat brain cortex synaptosomal fractions and, enriched in synaptosomal membranes according to marker analysis, were shown to have some phospholipase A activity, and as the synaptosomal junctions, as judged from morphological analysis, are rather resistant to phospholipase A, it is tempting to speculate on the possibility of synaptic areas containing such an enzyme, and through it interacting with localized areas in the cholinergic vesicle membrane containing the appropriate phospholipids. Specific alteration of the lipid components of the vesicular membrane could play a role in the release of ACh from synaptic vesicles. Adhesion of membranes was indicated in the observed cluster formation of vesicles upon addition of phospholipase A in early experiments. Such capacity to adhere could be a first step in a membrane fusion, a mechanism discussed, for example, in a recent paper by Poole *et al.* (1970). A temporary fusion of cholinergic vesicles with the junctional part of the external nerve ending membrane could be brought about by the presence of lysophosphatides formed in the membrane of the vesicle upon contact with a phospholipase A, that is either located in the synaptosomal membrane or being brought into contact with both membranes by, for example, an approaching lysosome. This could be followed by an emptying of the vesicle and its subsequent restoration by phospholipid-synthesizing enzymes present in the brain. This would uphold the observation that vesicles seem to survive the transmitter they carry. Further analysis will show if such enzymes are present in the external synaptosomal membrane. A permanent fusion of vesicle membrane and external synaptosomal membrane would be indicated if the lipids and proteins of both membranes were identical. From the limited evidence available (Whittaker, 1965) this is not the case. Permanent fusion would also demand that the external synaptosomal membrane creates new vesicles by some pinocytotic process, otherwise the membrane would grow. Though the origin of synaptic vesicles is not established, some current evidence points to the neurotubules or the cell soma rather than the external synaptosomal membrane. The membrane structure of neurotubules (Tilney and Porter, 1967) seems however to differ from that of the synaptic vesicles. Also, some morphological evidence (Brightman, 1967; Akert *et al.*, 1969) points to pinocytotic formation of synaptic vesicles from non-synaptic sites of the nerve terminal. Clearly, it is premature to discuss the origin of the cholinergic vesicle.

References pp. 38–39

SUMMARY

Synaptic vesicles, isolated from rat brain cortex, and cholinergic vesicles, isolated from the electric organ of *Torpedo nobiliana*, were broken down by a phospholipase A from cobra venom. The breakdown was accompanied by a release of acetylcholine. Morphological analysis revealed membrane fragments about one-third the size of the vesicle circumference. Subcellular fractions enriched in nerve ending membranes showed some phospholipase A activity. On the basis of these findings models of transmitter release, involving specific alterations of lipid components in the vesicular membrane, are discussed.

ACKNOWLEDGEMENTS

The skilled technical assistance of Ings Maj-Greth Nilsson, Kerstin Hermansson and C. De Laval is gratefully acknowledged.

REFERENCES

Abood, L. G. and Biel, J. H. (1962) Anticholinergic psychotomimetic agents. *Int. Rev. Neurobiol.*, **4**, 217–271.

Akert, K., Moor, H. Pfenninger, K. and Sandri, C. (1969) Contributions of new impregnation methods and freeze etching to the problems of synaptic fine structure. In *Prog. Brain Res.*, K. Akert and P. G. Waser (Eds.), *Vol. 31* Elsevier, Amsterdam, pp. 223–240.

Barker, L. A., Dowdall, M. J., Essman, W. B. and Whittaker, V. P. (1970) The compartmentation of acetylcholine in cholinergic nerve terminals. In *Drugs and Cholinergic Mechanisms in the CNS*, E. Heilbronn and A. Winter (Eds.), Res. Inst. of Natl. Defence, Stockholm, Almqvist and Wiksell, Stockholm, pp. 193–214.

Brightman, M. W. (1967) The intracerebral movement of proteins injected into blood and cerebrospinal fluid of mice. In *Prog. Brain Res.*, A. Lajtha and D. H. Ford (Eds.), *Vol. 29*, Elsevier, Amsterdam pp. 19–40.

Burton, R. M. and Gibbons, J. M. (1964) Lipid composition of a rat brain synaptic-vesicle fraction. *Biochim. Biophys. Acta. (Amst.)*, **84**, 220–223.

Cedergren, E., Heilbronn, E., Johansson, B. and Widlund, L. (1970) Ultrastructural stability of contact regions of phospholipase-treated synapses from rat motor cortex. *Brain Res.*, **24**, 139–142.

Condrea, E. and Rosenberg, P. (1968) Demonstration of phospholipid splitting as a factor responsible for increased permeability and block of axonal conduction induced by snake venom. II. Study on squid axons. *Biochim. biophys. Acta (Amst.)*, **150**, 271–284.

Durell, J. and Sodd, M. A. (1964) Studies on the acetylcholine-stimulated incorporation of radioactively labeled inorganic orthophosphate into the phospholipid of brain particulate preparations. *J. biol. Chem.*, **238**, 747–752.

Durell, J. and Sodd, M. A. (1966) Studies on the acetylcholine-stimulated incorporation of radioactive inorganic orthophosphate into the phospholipid of brain particulate preparations. II. Subcellular distribution of enzymic activity. *J. Neurochem.*, **13**, 487–491.

Ellman, G. E., Courtney, K. D., Andres, V. and Featherstone, R. M. (1961) A new and rapid colorimetric determination of acetylcholinesterase activity. *Biochem. Pharmacol.*, **7**, 88–95.

Gallai-Hatchard, J. and Gray, G. M. (1968) The action of phospholipase A on the plasma membrane of rat liver cells. *Europ. J. Biochem.*, **4**, 35–40.

Hanin, I. and Jenden, D. J. (1969) Estimation of choline esters in brain by new gas chromatographic preocedure. *Biochem. Pharmacol.*, **18**, 837–845.

Heilbronn, E. (1969) The effect of phospholipases on the uptake of atropine and acetylcholine by slices of mouse brain cortex. *J. Neurochem.*, **16**, 627–635.

HEILBRONN, E. AND CEDERGREN, E. (1970) Chemically induced changes in the acetylcholine uptake and storage capacity of brain tissue. In: E. HEILBRONN AND A. WINTER (Eds.), *Drugs and Cholinergic Mechanisms in the CNS*, Res. Inst. of Natl. Defence, Stockholm, Almqvist and Wiksell, Stockholm, pp. 245–265.

HEILBRONN, E. AND WIDLUND, L. (1970) The effect of *N*-methyl-4-piperidyl-diphenyl glycollate on the incorporation of ^{32}P into phospholipids from rat brain cortex slices and its subcellular localisation. *J. Neurochem.*, **17**, 1039–1048.

HEILBRONN, E., HAUSE, S. H. AND LUNDGREN, G. (1971) Chemical identification of acetylcholine in squid head ganglion. *Brain Res.*, **33**, 431–437.

HOKIN, L. E. AND HOKIN, M. R. (1955) Effects of acetylcholine on the phosphate turnover in phospholipids of brain cortex *in vitro*. *Biochim. biophys. Acta.*, **16**, 229–237.

HOKIN, L. E. AND HOKIN, M. R. (1958) Acetylcholine and the exchange of phosphate in phosphatidic acid in brain microsomes. *J. biol. Chem.*, **233**, 822–826.

LAPETINA, E. G. RODRIGUEZ DE LAES ARNAIZ, G. AND DE ROBERTIS, E. (1969) Turnover rates for glycerol, acetate, and orthophosphate in phospholipids of the rat cerebral cortex. *Biochim. biophys. Acta (Amst.)*, **176**, 643–646.

LARRABEE, M. G. (1968) Transynaptic stimulation of phosphatidylinositol metabolism in sympathetic neurons *in situ*. *J. Neurochem.*, **15**, 808.

LARRABEE, M. G., KLINGMAN, J. D. AND LEICHT, W. S. (1963) Effects of temperature, calcium and activity on phospholipid metabolism in a sympathetic ganglion. *J. Neurochem.*, **10** 549–570.

LARRABEE, M. G. AND LEITCH, W. S. (1965) Metabolism of phosphatidyl inositol and other lipids in active neurones of sympathetic ganglia and other peripheral nervous tissues. The site of the inositide effect. *J. Neurochem.*, **12**, 1–13.

LENARD, J. AND SINGER, S. J. (1968) Structure of membranes: reaction of red blood cell membranes with phospholipase C. *Science*, **159**, 738–739.

LOWRY, O. H., ROSENBROUGH, N. J., GARR, A. L. AND RANDALL, R. J. (1951) Protein measurement with the Folin phenol reagent. *J. biol. Chem.*, **193**, 265–275.

POOLE, A. R., HOWELL, J. I. AND LUCY, J. A. (1970) Lysolecithin and cell fusion. *Nature (Lond.)*, **227**, 810–814.

ROSENBURGH, P. AND CONDREA, E. (1968) Maintenance of axonal conduction and membrane permeability in the presence of extensive phospholipid splitting. *Biochem. Pharmacol.*, **17**, 2033–2044.

TILNEY, L. G. AND PORTER, K. R. (1967) Studies on the microtubules in Heliozoa. II. The effect of low temperature on these structures in the formation and maintenance of axopodia. *J. Cell. Biol.*, **34**, 327–343.

VON HUNGEN, K., MAHLER, R. AND MOORE, W. J. (1968) Turnover of protein and ribonucleic acid in synaptic subcellular fractions from rat brain. *J. biol. Chem.*, **243**, 1415–1423.

WEBSTER, G. R. (1965) The acylation of lysophosphatides with longchain fatty acids by rat brain and other tissues. *Biochim. biophys. Acta (Amst.)*, **98**, 512–519.

WHITTAKER, V. P. (1959) The isolation and characterisation of acetylcholine-containing particles from brain. *Biochem. J.*, **72**, 694–706.

WHITTAKER, V. P. (1965) The application of subcellular fractionation techniques to the study of brain function. *Prog. Biophys.*, **15**, 39–96.

DISCUSSION

LEADBEATER: Would you care to expand your remarks on the effects of glycollates, and sodium and calcium ions on the uptake of ACh into brain slices? What significance do these observations have concerning the mode of action of the glycollates in producing psychotomimetic effects?

HEILBRONN: Pharmacologists think of glycollates as acting like atropine on the postsynaptic receptor. It seems obvious to me that there is a presynaptic action as well which has to do with the release of ACh. There is enough evidence in the literature to show that glycollates and atropine release ACh perhaps by interfering with an inhibitory mechanism.

ANSELL: Have you investigated which phospholipids in the synaptic vesicle membrane are broken down by the *Naja Naja* phospholipase A?

Reference p. 40

comparably simple, layered structure.* The efferent neurones of the region, the pyramidal cells, are located in a discrete layer extending throughout the hippocampus (Fig. 1). These cells are inhibited by nerve terminals on their cell bodies (Andersen *et al.*, 1964) and excited by terminals on their dendrites (Andersen *et al.*, 1966). The inhibitory terminals belong to basket cells situated mainly below the pyramids in the stratum oriens. The excitatory terminals derive from various sources and are distributed in a layered manner (for review of hippocampal anatomy see Blackstad, 1967). An analogous situation exists for the granular cells in area dentata (Andersen *et al.*, 1966). In the hippocampal region different kinds of nerve terminals are thus spatially separated and samples enriched in a particular type of terminal can be obtained by microdissection.

ACETYLCHOLINE

The putative transmitter acetylcholine (ACh) is synthesized by choline acetyltransferase (ChAc) and hydrolysed by acetylcholinesterase (AChE). ACh is relatively labile in the tissue and at present it cannot be determined in small tissue samples. The level of ACh is, however, governed by ChAc, which is therefore a reliable marker for cholinergic structures (Hebb, 1963; Silver, 1967). AChE is less reliable as a cholinergic marker, but has the advantage that it is easily demonstrated by histochemical staining, which allows a detailed mapping of its localization.

Histochemical staining for AChE reveals a striking laminar distribution in the hippocampal region (Fig. 2) (Storm–Mathisen and Blackstad, 1964). We made use of this laminar distribution of AChE in quantitative histochemical studies of AChE (Storm–Mathisen, 1970) and ChAc (Fonnum, 1970). From freeze-dried sections we dissected the zones of hippocampus regio superior and area dentata displayed by histochemical staining. Samples weighing 0.05–2 μg were assayed by microchemical procedures (Fonnum 1969). The distribution of ChAc and AChE, measured biochemically (Figs. 3 and 4; Table I), was very similar and in excellent agreement with the histochemical staining pattern (Fig. 2). When the activities were expressed per tissue volume they correlated closely with the staining intensities microdensitometrically measured in sections treated by the Koelle method (Storm–Mathisen, 1970).

The distribution pattern did not strictly conform to that of any of the anatomically described afferent fibre systems. The maximum enzyme activity was found in a zone immediately subjacent to the pyramidal cells and in the two zones adjacent to the granular cells (Figs. 2, 3 and 4). Like these zones the interstices between the pyramidal and granular cells were heavily stained, whereas the cell bodies were pale. Correspondingly, fine dissection of the area around the pyramidal cells showed a reduced activity at the level of the cell bodies (Fig. 4). The zones receiving commissural fibres,

* The hippocampal region comprises area dentata, hippocampus, subiculum, presubiculum, area retrosplenialis, parasubiculum and area entorhinalis. In this paper only the two former parts are dealt with. For nomenclature see Blackstad (1956).

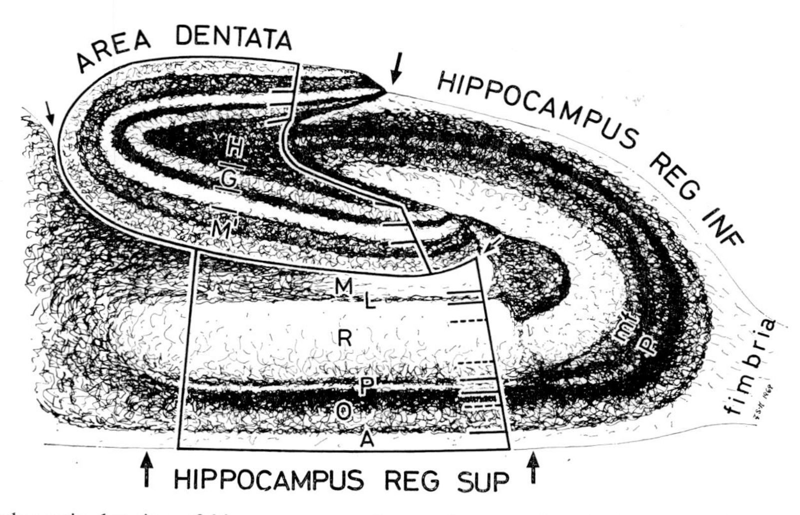

Fig. 2. Schematic drawing of hippocampus and area dentata of rat brain as seen in a horizontal section stained for AChE. The darkness of the various zones indicates their activities of AChE and ChAc (Storm–Mathisen, 1970; Fonnum, 1970). The nomenclature is according to Blackstad (1956). Small arrows point to the obliterated fissura hippocampi, which represents the surfaces of the hippocampal and dentate cortices. The areas from which samples were dissected are circumscribed by solid lines. The zones dissected from regio superior (REG SUP) were: A, alveus; O, stratum oriens (divided into Oi and Oo); P, stratum pyramidale including 10–20 μm of each of the adjacent layers; R, stratum radiatum (divided into Ri, Rm and Ro); L, stratum lacunosum; M, stratum moleculare. From area dentata the following zones were dissected for GAD assay (Fig. 6): M′, outer two-thirds of stratum moleculare (the inner one-third, containing terminals of commissural fibres, was discarded); G, stratum granulare plus 20–30 μm of each of the adjacent AChE-rich zones; H, hilus fasciae dentatae. For assay of ChAc (Fig. 3) and AChE, M′ was subdivided into Mo and Mm and the underlying parts of area dentata were subdivided into 4 zones with limits differing from the ones in this figure, but with widths as indicated in Fig. 3 and with the center of zone G coinciding in the two figures. (Reproduced from *J. Neurochem.* (1971), Storm–Mathisen and Fonnum, **18**, 1105–1111.)

i.e., axons from pyramidal cells in the contralateral hippocampus, were either poor (stratum radiatum of regio superior, zone in inner third of molecular layer in area dentata) or rich (stratum oriens) in ChAc and AChE. The zones containing the nerve terminals of the afferents from area entorhinalis (Fig. 7A, perforant path) and of the axons of the granular cells (Fig. 7D, mossy fibres) contained mostly moderate amounts.

In an electron microscope study Shute and Lewis (1966) found most of the AChE to be presynaptic. Scattered local neurones (Golgi type II cells) in hilus fasciae dentatae and stratum oriens stained for AChE. Terminals on the bodies of the pyramidal and granular cells, *i.e.*, the inhibitory terminals of the basket cells, were as a rule devoid of AChE. Further, they observed that AChE-positive nerve terminals were less numerous than AChE-positive thin nerve fibres, and concluded that the pattern of AChE staining reflected the distribution of an AChE-containing preterminal nerve plexus rather than that of AChE-containing boutons.

However, subcellular fractionation of the hippocampal region (Fonnum, 1970) showed that ChAc was strongly concentrated in synaptosomes (Tables II and III). By far the greater part was found in the heavier fraction, P_2H (Table III). The thin preterminal nerve fibres (0.1–0.2 μm), interpreted by Shute and Lewis to contain

References pp. 55–57

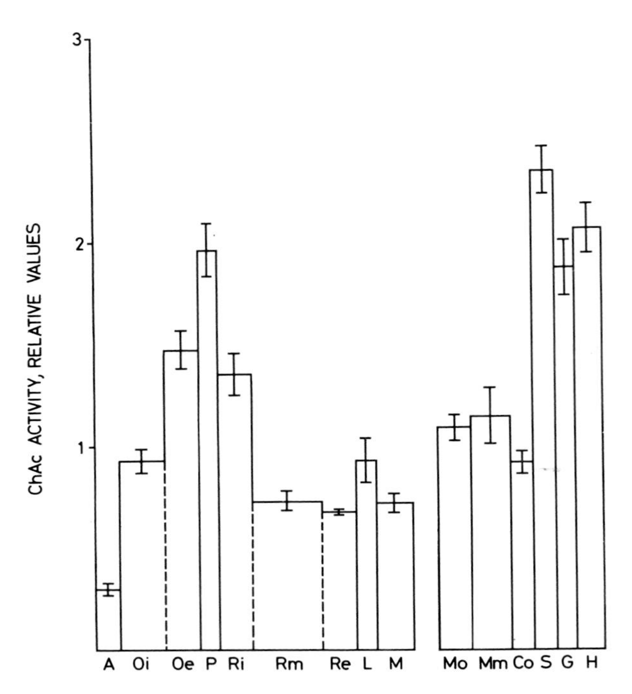

Fig. 3. Distribution of ChAc in different zones of hippocampus regio superior and area dentata. The columns represent mean activities and the top bars equal twice S.E.M. For each zone the ChAc activity is given relative to that of strips cut at right angles through all layers of hippocampus regio superior. This was 21.7 ± 0.8 μmoles/h/g dry wt. (mean ± S.E.M., 7 animals) for the series from hippocampus and 26.3 ± 3.3 μmoles/h/g dry wt. (mean ± S.E.M., 3 animals) for area dentata. The abscissa gives approximate widths of dissected samples, the total width through all layers of hippocampus regio superior being *ca.* 800 μm. For dissection, see Fig. 2. The number of samples/number of animals for each zone was: A 4/4, Oi 8/6, Oe 9/7, P 7/5, Ri 3/2, Rm 9/6, Ro 4/3, L 3/2, M 5/5, Mo 9/3, Mm 9/3, Co 11/3, S 11/3, G 7/3 and H 8/3. The data presented for zone M were from its rostral (pale) part. The caudal part had 1.3–1.4 times higher activity. (Redrawn from *J. Neurochem.* (1970), F. Fonnum, **17**, 1029–1034.)

most of the AChE, would be expected to form small particles sedimenting in the lighter fractions P_2L' and P_3. Since quantitative differences in enzyme activity are not easily revealed in electron microscopic histochemistry it is conceivable that a relatively large area of stained axon membranes could represent only a minor part of the total enzyme activity. It may also be difficult to distinguish between preterminal axons and boutons in histochemical material, especially when serial sections are not used. The electron microscope observations therefore do not necessarily contradict the finding that the greater part of ChAc and AChE is present in nerve terminals.

A very substantial contribution to the identification of the cholinergic nerve elements in the hippocampal region was the demonstration by Shute and Lewis (1961) and Lewis and Shute (1967) that AChE disappears from the neuropil of this region after transection of the fimbria (Fig. 5). Corresponding results were obtained for ChAc (Lewis *et al.*, 1967). The few perikarya that stained for AChE, retained their staining after the transection. AChE and ChAc accumulated in fibres in fimbria

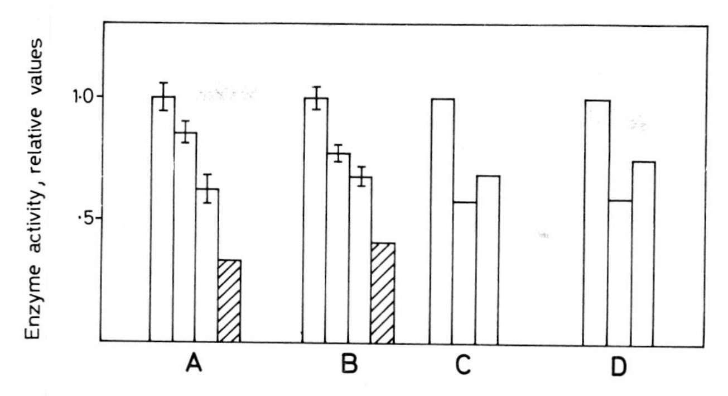

Fig. 4. Detailed distribution of ChAc and AChE in the vicinity of the pyramidal cells. White columns denote mean activities of 20–30 μm wide zones of hippocampus regio superior representing stratum pyramidale (middle), the adjacent zone of stratum oriens (left) and of stratum radiation (right). Hatched columns represent mean activities of strips cut at right angles through all layers of regio superior. All activities are presented relative to the highest in each group. Top bars equal twice S.E.M. A, ChAc activities expressed per dry weight. B, AChE activities per dry weight. C, AChE activities per volume. D, Microdensitometric measurements on section stained for AChE by the Koelle method. Note that the highest activities are in the infrapyramidal zone and that the biochemically measured activities correlate well with the staining results. (Data from *J. Neurochem.* (1970), (A) F. Fonnum, **17**, 1029–1037, (B, C and D) J. Storm-Mathisen, **17**, 739–750).

TABLE I

COMPARISON OF ChAc AND AChE ACTIVITIES IN LAYERS OF HIPPOCAMPUS REGIO SUPERIOR AND AREA DENTATA

	Hippocampus reg. sup.				*Area dentata*		
Zone	*AChE*	*ChAc*	*Ratio AChE/ChAc*	*Zone*	*AChE*	*ChAc*	*Ratio AChE/ChAc*
A	550	6·5	85	Mo	2280	28·5	80
Oi	1960	19·9	98	Mm	2610	30·1	87
Oo	2780	31·2	89	Co	1910	23·6	81
P	3730	43·8	85	S	4530	61·7	74
Ri	2540	28·7	89	G	3380	48·9	69
Rm	1570	15·8	99	H	4140	53·7	77
Ro	1030	15·6	66	All layers area dentata	3070	40·6	77
L	1480	20·4	74				
M	1160	15·1	78				
All layers reg. sup.	1750	21·7	81	All layers hippocampus reg. sup.	1890	26·3	76

The areas dissected and the nomenclature are shown in Fig. 2. The results are mean values of 4–12 measurements from 1 to 7 different animals. ChAc values are expressed as μmoles ACh synthesized/h/g dry wt. and are taken from Fonnum (1970). AChE values are from Storm-Mathisen (1970) and are expressed as μmoles acetylthiocholine hydrolysed/h/g dry wt. (hippocampus regio superior) or as μmoles ACh hydrolysed/h/g dry wt. (area dentata).

References pp. 55–57

TABLE II

SUBCELLULAR DISTRIBUTION OF ChAc, LACTATE DEHYDROGENASE AND PROTEIN IN RAT HIPPOCAMPAL REGION

Fraction	Distribution (per cent of total recovered material)			Relative specific concentration	
	ChAc	*Lactate dehydrogenase*	*Protein*	*ChAc*	*Lactate dehydrogenase*
P_1 (crude nuclear)	14·1	12·2	23·0	0·61	0·53
P_2 (crude mitochondrial)	62·3	42·6	47·0	1·32	0·91
A (myelin)	7·2	6·0	9·4	0·75	0·65
B (synaptosomal)	48·7	25·2	23·5	2·07	1·07
C (mitochondrial)	6·4	11·4	14·1	0·38	0·81
S_2 (supernatant)	23·5	45·2	30·0	0·78	1·51

The results represent mean values of 3 experiments. Fractions were prepared as described by Gray and Whittaker (1962). The recoveries of the components measured were between 90 and 110%. $P_1 + P_2 + S_2 = 100\%$, $P_2 = A + B + C$. The RSA is the percentage of a compound in any fraction divided by the percentage of protein in the same fraction. The concentrations of enzyme activities and proteins in hippocampus homogenate were: ChAc, 7.8 μmoles/h/g; lactate dehydrogenase, 47 μmoles/min/g; protein, 95 mg/g. (From Fonnum, 1970)

TABLE III

SUBCELLULAR DISTRIBUTION OF ChAc AND AChE IN RAT HIPPOCAMPAL REGION

Fraction	Distribution (per cent of total recovered material)			Ratio
	AChE	*ChAc*	*Protein*	*AChE/ChAc*
P_1'	20·0	21·3	30·9	0·9
P_2H'	47·1	46·4	36·1	1·0
P_2L'	23·9	16·6	10·1	1·4
P_3'	4·5	1·0	2·4	4·5
S_3'	4·6	14·5	21·5	0·3

The results represent mean values of 2 experiments. The fractions were prepared as follows: P_1' (1000 g × 10 min, washed once); P_2H' (6000 g × 20 min, "heavy synaptosomes"); P_2L' (17000 g × 60 min, "light synaptosomes "and membrane fragments); P_3 (100000 g × 60 min); S_3' (final supernatant). The AChE activity in the homogenate was 8.4 μmoles/min/g. (From Fonnum, 1970)

anterior to the lesion. The fibres probably arise from small cells staining heavily for AChE and situated in the medial septal nucleus and in the nucleus of the diagonal band (Lewis and Shute, 1967). In agreement, McGeer *et al.* (1969) found a singnificant decline of ChAc in the hippocampal region after lesions involving the septum.

Combining experimental neuroanatomical techniques with "quantitative histochemistry" we made the following study (Storm–Mathisen, 1972). The afferents from the area entorhinalis (perforant path), the fimbria (from septum, columna fornicis and contralateral hippocampus) and the "intrinsic" fibres from area dentata to

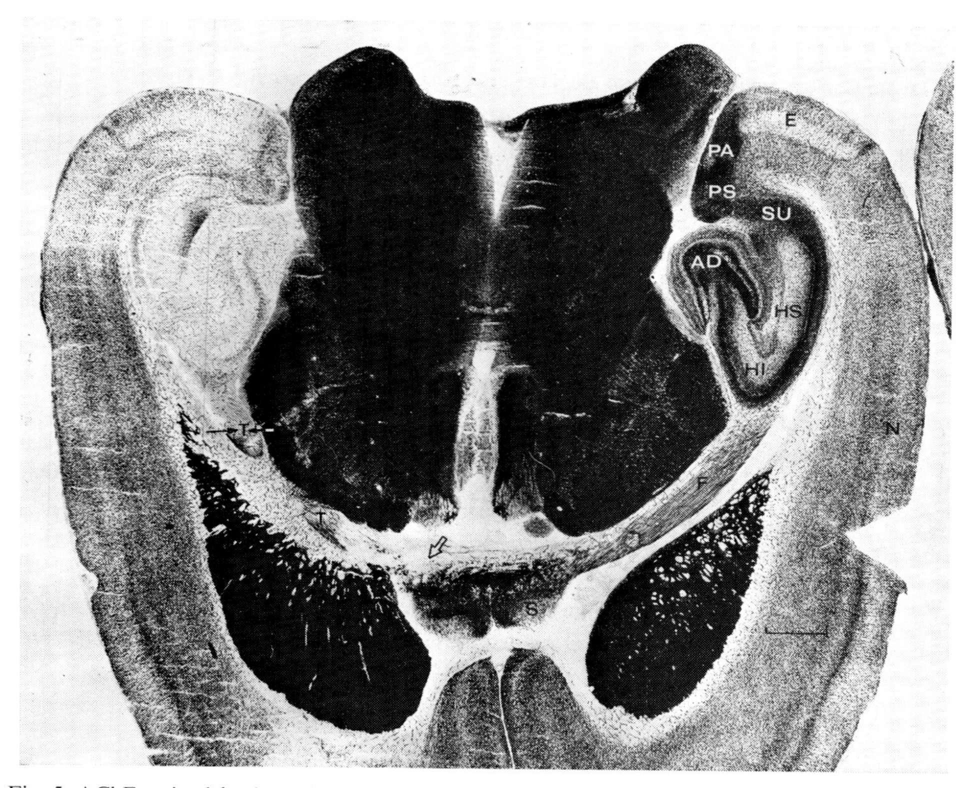

Fig. 5. AChE-stained horizontal section of a rat brain 7 days after transection of fimbria (coarse arrow). Note massive reduction of staining intensity and loss of laminar staining pattern on operated side. Capillaries stain equally well on both sides (non-specific cholinesterase was not inhibited in this section). Small arrows point to limit between normally staining (intact) stria terminalis and pale fimbria on the operated side. S, septum; F, fimbria; AD, area dentata; HI, hippocampus regio inferior; HS, hippocampus regio superior; SU, subiculum; PS, presubiculum; PA, parasubiculum; E, area entorhinalis; T, stria terminalis; N, neocortex. Calibration bar 1 mm.

hippocampus regio inferior (mossy fibres) were interrupted surgically in rats. After various survival times tissue samples were obtained and analysed as previously described. In addition, regularly spaced sections were mounted, fixed and stained for degenerating nerve terminals (Hjorth–Simonsen, 1970) and for AChE.

Transection of fimbria led to an almost complete loss of ChAc (as well as of AChE) from all the zones on the operated side (Table IV), whereas there was no change on the "normal" side. The shortest survival time tested (4.5 days) gave the same loss of activity as the longest. None of the other types of lesion had any significant effect (Table V). Such results virtually exclude the possibility that the cholinergic fibres could belong to any of the above mentioned afferent systems, or to the local basket cell plexuses, although the latter have a distribution somewhat similar to that of ChAc and AChE. It is unlikely that the changes in enzyme activities could be ascribed to transneuronal effects (for discussion see Storm–Mathisen, 1970).

The above results therefore support the view of Lewis and Shute (1967) that in the

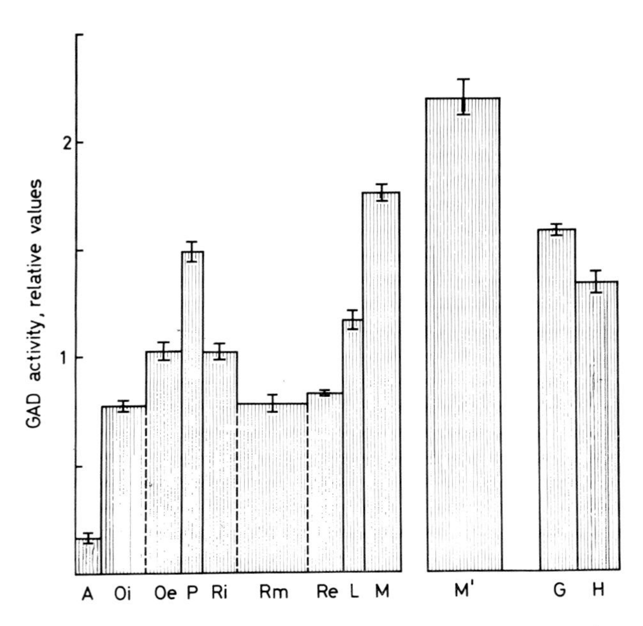

Fig. 6. Distribution of GAD in different zones of hippocampus regio superior and area dentata. Presentation as in Fig. 3, dissection described in Fig. 2. The mean activity of samples cut through all layers of hippocampus regio superior (activity unit) was in the series from hippocampus 44.5 ± 4.7 μmoles/h/g dry wt. (S.D., 18 samples) and in that from area dentata 44.0 ± 4.0 μmoles/h/g dry wt. (S.D., 8 samples). The data were from two animals and the numbers of samples for the different zones were: A 10, Oi 7, Oe 7, P 7, Ri 6, Rm 10, Re 4, L 7, M 11, M′ 8, G 9 and H 8. (Reproduced from *J. Neurochem.* (1971), Storm–Mathisen and Fonnum, **18**, 1105–1111.)

and in the adjoining zone of stratum radiatum, which had similar activities. The former zone contains the inhibitory basket cell terminals whereas the latter two contain excitatory terminals. In contrast, the activity of ChAc in the mossy fibre layer was nearly twice that in stratum radiatum, and only slightly lower than that in the pyramidal zone. In area dentata, as in hippocampus regio superior, the molecular layer had a higher activity than the cell layer. This was again higher than in the hilus, and showed about the same activity as the pyramidal layers (Fig. 6). In a separate experiment the outer third of the molecular layer was found to have 65 % higher activity than the middle third.

In fimbria, the activity of GAD was low compared to that of ChAc (Storm–Mathisen and Fonnum, 1971). This suggested that GAD would reside in afferents arriving through another pathway, or in intrahippocampal neurones. This hypothesis was tested in the material used for ChAc, where the anatomically described afferent nervous pathways were surgically interrupted (see above) (Storm–Mathisen, 1972). Samples were dissected from the levels showing maximum degeneration of nerve terminals (Fig. 7). Nerve terminal particles contain most of the GAD activity in the cerebral cortex (Fonnum, 1968). None of the lesions resulted in any important reduction of GAD (Tables IV and VI). The fimbrial lesions gave a slight reduction of a maximum 16 %. This could conceivably be due to "trophic" changes after removal of a major part of the afferents (commissural and septal fibres and fibres ascending in the fornix).

It may be concluded that nearly all, or possibly all, GAD activity in the hippocampal region is present in local neurones. Therefore, the fact that granular and pyramidal cell layers, which contain the inhibitory terminals of local neurones, *viz.* the basket cells, showed a uniformly high activity (Fig. 6; Table V) supports the view that GAD is localized in these terminals.

The very high activity found in the molecular layers cannot be interpreted at present, but it may be noted that these layers contain local neurones morphologically somewhat similar to the basket cells (Ramón y Cajal, 1893; Lorente de Nó, 1934), but with unknown function.

GABA has been found to inhibit pyramidal cells in the hippocampus (Biscoe and Straughan, 1966; Salmoiraghi and Stefanis, 1967). Synaptic inhibition of these cells is blocked by bicuculline, a specific GABA antagonist, but not by strychnine, which does not act on GABA synapses (Curtis *et al.*, 1970). The inhibitory synapses on the granular cells in area dentata are also strychnine insensitive (Andersen *et al.*, 1966). GABA has not been shown to be released from the hippocampal cortex during inhibitory synaptic activity. This has been done in the neocortex (Mitchell and Srinivasan, 1969; Iversen *et al.*, 1970), where microiontophoretically applied GABA also has been demonstrated to mimic the inhibitory postsynaptic potentials (Krnjević and Schwartz, 1967).

The evidence accumulated so far strongly favours GABA as the inhibitory transmitter of the basket cells.

OTHER TRANSMITTER CANDIDATES

The data on other transmitter candidates in the hippocampus are less complete. All parts of the hippocampal region contain noradrenaline (NA) terminals which are, however, diffusely distributed relative to ChAc and constitute but a small proportion of the afferents (Blackstad *et al.*, 1967). The highest levels are found in hilus of area dentata and in stratum radiatum of regio inferior. The distribution of nerve terminals containing serotonin (5-HT) has not been described, apart from stating that it is different from that of NA terminals (Fuxe *et al.*, 1970), probably because of the instability of the fluorescence obtained with this substance in the histochemical technique. However, there has been a report of 5-HT terminals on the pyramidal cell bodies in the cat (Eidelberg *et al.*, 1967).

The NA and 5-HT fibres arise from NA- and 5-HT-containing neurones in the brain stem. They have been described as ascending in the medial forebrain bundle and reaching the hippocampus through fimbria and fornix superior (Fuxe, 1965). After lesions involving the medial forebrain bundle, 5-HT decreased by 68% in the hippocampus (Moore and Heller, 1967), and NA decreased by 71% in the hippocampus and amygdala (Heller *et al.*, 1966). We observed an almost total (about 90%) loss of 5-HT from the hippocampus on the operated side 6 days after unilateral trans-

References pp. 55–57

action of fimbria, fornix superior and the dorsal hippocampus (Storm–Mathisen and Guldberg, unpublished observations). McGeer *et al.* (1969) found a loss of tryptophan 5-hydroxylase and, less consistently, of tyrosine hydroxylase from the hippocampus after lesions in the septal area.

These results indicate that 5-HT and NA in the hippocampal region are not localized in intrinsic neurones, but in terminals of afferent nerve fibres, probably arriving *via* the medial forebrain bundle from cell bodies in the brain stem.

According to the pharmacological data the predominant effects of 5-HT and NA in the hippocampus seem to be inhibitory (Herz and Nacimiento, 1965; Biscoe and Straughan, 1966; Eidelberg *et al.*, 1967; Salmoiraghi and Stefanis, 1967). Release experiments have not been reported.

Finally the interesting observation of a remarkable concentration of zinc in the excitatory mossy fibre system should be mentioned (Maske, 1955). The metal is localized in the mossy fibre boutons and electron microscope observations indicate that it is associated with the synaptic vesicles (Haug *et al.*, 1971, and references therein). Synaptic transmission in the mossy fibre system is selectively blocked by the application of hydrogen sulfide (Euler, 1962), indicating that zinc may be involved in the transmission process.

CONCLUSION

Our present state of knowledge of transmitters in the hippocampal region is then as follows. The cholinergic elements of the region are nerve terminals concentrated mainly in an infrapyramidal zone in hippocampus and in the hilus as well as in a supragranular band of area dentata. These terminals arise from afferents arriving through the fimbria and superior fornix from cells in the medial septal nucleus and the nucleus of the diagonal band. They are probably excitatory.

The GABA-producing neurones are situated entirely within the hippocampal region. Among these are most likely the basket cells, which produce recurrent inhibition, seemingly by releasing GABA onto the bodies of pyramidal and granular cells.

NA is present in relatively diffusely distributed nerve terminals of afferents passing through the fimbria and superior fornix, and probably ascending with the medial forebrain bundle from cells in the brain stem. It is likely that these terminals are inhibitory.

5-HT is contained in nerve terminals which arise from separate groups of brain stem neurones and have another distribution than that of the NA terminals, but otherwise show the same features.

SUMMARY

In the hippocampal region different types of nerve terminals are localized in different zones and it is therefore possible to correlate the localization of particular transmitter candidates with the localization of particular types of nerve terminals.

The localization of ChAc and GAD was investigated by quantitative histochemical methods in normal animals and in animals with lesions of afferent nerve pathways. In addition the subcellular distribution of ChAc was studied. ChAc is concentrated in nerve terminals localized mainly in a narrow layer on both sides of the somata of the granular and pyramidal cells and derived from cells in the medial part of the septum. GAD is localized in intrinsic neurones and its distribution suggests that part of it is concentrated in the inhibitory basket cells.

The literature on ACh, GABA and aromatic monoamines as potential transmitters in the hippocampal region is reviewed.

ACKNOWLEDGEMENTS

We are grateful to Mrs. E. Iversen, Mr. J. E. Frogner, Mr. H. A. T. Saxholm, Mrs. B. Branil and Mrs. E. Rieber–Mohn for skilled technical assistance. Most of the operated animals were prepared in the Anatomical Institute, University of Aarhus, Denmark. We wish to thank Professor T. W. Blackstad and the other members of the staff for their hospitality and stimulating interest. We are also indebted to Professor F. Walberg and the other members of the staff at the Anatomical Institute, University of Oslo, Norway, for their never-failing helpfulness and advice.

Figures 2, 3, 4 and 6, and Tables 1, 2 and 3 were reprinted by permission of the *Journal of Neurochemistry*.

REFERENCES

Alksne, J. F., Blackstad, T. W., Walberg, F. and White, L. E. Jr. (1966) Electron microscopy of axon degeneration: A valuable tool in experimental neuroanatomy. *Ergebn. Anat. Entwickl.-Gesch.*, **39**, 3–32.

Andersen, P., Blackstad, T. W. and Lømo, T. (1966) Location and identification of excitatory synapses on hippocampal pyramidal cells. *Exp. Brain Res.*, *1*, 236–248.

Andersen, P., Bruland, H. and Kaada, B. R. (1961a) Activation of the dentate area by septal stimulation. *Acta physiol. scand.*, **51**, 17–28.

Andersen, P., Bruland, H. and Kaada, B. R. (1961 b) Activation of field CA1 of the hippocampus by septal stimulation. *Acta physiol. scand.*, **51**, 29–40.

Andersen, P., Eccles, J. C. and Løyning, Y. (1964) Pathway of postsynaptic inhibition in the hippocampus. *J. Neurophysiol.*, **27**, 608–619.

Andersen, P., Holmqvist, B. and Voorhoeve, P. E. (1966) Entorhinal activation of dentate granule cells. *Acta physiol. scand.*, **66**, 448–460.

Baxter, C. F. (1970) The nature of γ-aminobutyric acid. In *Handbook of Neurochemistry*, A. Lajtha (Ed.), Vol. III, Plenum Press, New York, pp. 289–353.

Biscoe, T. J. and Straughan, D. W. (1966) Microelectrophoretic studies of neurones in the cat hippocampus. *J. Physiol. (Lond.)*, **183**, 341–359.

Blackstad, T. W. (1956) Commissural connections of the hippocampal region in the rat, with special reference to their mode of termination. *J. comp. Neurol.*, **105**, 417–537.

Blackstad, T. W. (1967) Cortical gray matter. A correlation of light and electron microscopic data. In *The Neuron*, H. Hydén (Ed.), Elsevier, Amsterdam, pp. 49–118.

Blackstad, T. W., Fuxe, K. and Hökfelt, T. (1967) Noradrenaline nerve terminals in the hippocampal region of the rat and the guinea pig. *Z. Zellforsch.*, **78**, 463–473.

Brücke, F., Gogolák, G. and Stumpf, C. (1963) Mikroelektrodenuntersuchung der Reitzantwort und der Zelltätigkeit im Hippocampus bei Septumreizung. *Arch. ges. Physiol.*, **276**, 456–470.

COLLIER, B. AND MITCHELL, J. F. (1967) The central release of acetylcholine during consciousness and after brain lesions. *J. Physiol. (Lond.)*, **188**, 83–98.

CURTIS, D. R., FELIX, D. AND MCLENNAN, H. (1970) GABA and hippocampal inhibition. *Brit. J. Pharmacol.*, **40**, 881–883.

EIDELBERG, E., GOLDSTEIN, G. P. AND DEZA, L. (1967) Evidence for serotonin as a possible inhibitory transmitter in some limbic structures. *Exp. Brain Res.*, **4**, 73–80.

EULER, C. VON (1962) On the significance of the high zinc content in the hippocampal formation. In *Physiologie de L'Hippocampe*, Editions du Centre National de la Recherche Scientifique, Paris, pp. 135–145.

FONNUM, F. (1968) The distribution of glutamate decarboxylase and aspartate transaminase in subcellular fractions of rat and guinea-pig brain. *Biochem. J.*, **106**, 401–412.

FONNUM, F. (1969) Radiochemical micro-assays for the determination of choline acetyltransferase and acetylcholinesterase activities. *Biochem. J.*, **115**, 465–472.

FONNUM, F. (1970) Topographical and subcellular localization of choline acetyltransferase in rat hippocampal region. *J. Neurochem.*, **17**, 1029–1037.

FUXE, K. (1965) Evidence for the existence of monoamine neurons in the central nervous system. IV. Distribution of monoamine nerve terminals in the central nervous system. *Acta physiol. scand.*, Suppl. 247, **64**, 37–84.

FUXE, K., HÖKFELT, T., JONSSON, G. AND UNGERSTEDT, V. (1970) Fluorescence microscopy in neuro-anatomy. In *Contemporary Research Methods in Neuroanatomy*, V. J. H. NAUTA AND S. O. E. EBBESSON (Eds.), Springer, Berlin, pp. 275–314.

GRAY, E. G. AND WHITTAKER, V. P. (1962) The isolation of nerve endings from brain: an electron microscopic study of cell fragments derived by homogenization and centrifugation. *J. Anat. (Lond.)*, **96**, 79–88.

HAUG, F.-M. Š., BLACKSTAD, T. W., HJORTH-SIMONSEN, A. AND ZIMMER, J. (1971) Timm's sulfide silver reaction for zinc during experimental anterograde degeneration of hippocampal mossy fibres. *J. comp. Neurol.*, **142**, 23–32.

HEBB, C. (1963) Formation, storage and liberation of acetylcholine. In *Cholinesterases and Anti-cholinesterase Agents*, G. B. KOELLE (Ed.), Springer, Berlin, pp. 55–88.

HELLER, A., SEIDEN, L. S. AND MOORE, R. Y. (1966) Regional effects of lateral hypothalamic lesions on brain norepinephrine in the cat. *Int. J. Neuropharmacol.* **5**, 91–101.

HERZ, A. AND NACIMIENTO, A. C. (1965) Über die Wirkung von Pharmaka auf Neurone des Hippocampus nach mikroelektrophoretischer Verabfolgung. *Arch. exp. Path. Pharmak.*, **251**, 295–314.

HJORTH-SIMONSEN, A. (1970) Fink–Heimer silver impregnation of degenerating axons and terminals in mounted cryostate sections of fresh and fixed brains. *Stain Technol.*, **45**, 199–204.

HJORTH-SIMONSEN, A. (1972) Projection of the lateral part of the entorhinal area to the hippocampus and fascia dentata. *J. comp. Neurol.*, in press.

IVERSEN, L. L., MITCHELL, J. F., NEAL, M. J. AND SRINIVASAN, V. (1970) The effect of stimulation of inhibitory pathways on the release of endogenous γ-aminobutyric acid from cat cerebral cortex. *Brit. J. Pharmacol.*, **38**, 452P.

KRNJEVIĆ, K. AND SCHWARTZ, S. (1967) The action of γ-aminobutyric acid on cortical neurones. *Exp. Brain Res.*, **3**, 320–336.

LEWIS, P. R. AND SHUTE, C. C. D. (1967) The cholinergic limbic system: Projections to hippocampal formation, medial cortex, nuclei of the ascending cholinergic reticular system, and the sub-fornical organ and supraoptic crest. *Brain*, **90**, 521–540.

LEWIS, P. R., SHUTE, C. C. D. AND SILVER, A. (1967) Confirmation from choline acetylase analyses of a massive cholinergic innervation to the rat hippocampus. *J. Physiol. (Lond.)*, **191**, 215–224.

LORENTE DE NÓ, R. (1934) Studies on the structure of the cerebral cortex. II. Continuation of the study of the ammonic system. *J. Psychol. Neurol. (Lpz.)*, **46**, 113–177.

MASKE, H. (1955) Über den topochemischen Nachweis von Zinc im Ammonshorn verschiedener Säugetiere. *Naturwissenschaften*, **42**, 424.

MCGEER, E. G., WADA, J. A., TERAO, A. AND JUNG, E. (1969) Amine synthesis in various brain regions with caudate or septal lesions. *Exp. Neurol.*, **24**, 277–284.

MITCHELL, J. F. AND SRINIVASAN, V. (1969) Release of ^{3}H-γ-aminobutyric acid from the brain during synaptic inhibition. *Nature (Lond.)*, **224**, 663–666.

MOORE, R. Y. AND HELLER, A. (1967) Monoamine levels and neuronal degeneration in rat brain following lateral hypothalamic lesions. *J. Pharmacol. exp. Ther.*, **156**, 12–22.

RAISMAN, G. (1966) The connexions of the septum. *Brain*, **89**, 317–348.

RAMÓN Y CAJAL, S. (1893) *The Structure of Ammon's Horn.* (Trans. by L. M. KRAFT, 1968) Thomas, Springfield, Ill.
SALMOIRAGHI, G. C. AND STEFANIS, C. N. (1967) A critique of iontophoretic studies of central nervous system neurons. *Int. Rev. Neurobiol.*, **10**, 1–30.
SHUTE, C. C. D. AND LEWIS, P. R. (1961) The use of cholinesterase techniques combined with operative procedures to follow nervous pathways in the brain. *Bibl. anat. (Basel)*, **2**, 34–49.
SHUTE, C. C. D. AND LEWIS, P. R. (1966) Electron microscopy of cholinergic terminals and acetylcholinesterase-containing neurones in the hippocampal formation of the rat. *Z.Zellforsch.*, **69**, 334–343.
SILVER, A. (1967) Cholinesterase of the central nervous tissue with special reference to the cerebellum. *Int. Rev. Neurobiol.*, **10**, 57–109.
STORM–MATHISEN, J. (1970) Quantitative histochemistry of acetyl-cholinesterase in rat hippocampal region correlated to histochemical staining. *J. Neurochem.*, **17**, 739–750.
STORM-MATHISEN, J. (1972) Glutamate decarboxylase in the rat hippocampal region after lesions of the afferent fibre systems. Evidence that the enzyme is localised in intrinsic neurones. *Brain Res.*, in press.
STORM–MATHISEN, J. AND BLACKSTAD, T. W. (1964) Cholinesterase in the hippocampal region. Distribution and relation to architectonics and afferent systems. *Acta anat. (Basel)*, **56**, 216–253.
STORM–MATHISEN, J. AND FONNUM, F. (1971) Quantitative histochemistry of glutamate decarboxylase in the rat hippocampal region. *J. Neurochem.*, **18**, 1105–1111.

DISCUSSION

BRIMBLECOMBE: Have you any idea at all concerning the functional significance of the zinc which is localized in the hippocampus and are there similar localizations in any other areas of the brain?

STORM–MATHISEN: Its high concentration in the mossy fibres is unique and it seems well established that it is zinc by radio-chemical substitution and by atomic absorption analysis. Perhaps it has something to do with binding of a transmitter or perhaps it acts by itself.

BRIMBLECOMBE: Is it found elsewhere in the brain?

STORM–MATHISEN: Not at that high concentration, but it is found elsewhere and it has a very conspicuous laminar distribution. (F. M. Š. Haug, unpublished observation).

HOWELLS: Could you clarify the distribution of GAD and ChAc in the hippocampus? Was the distribution of GAD parallel to that of ChAc?

STORM–MATHISEN: No, they were not parallel except, it seems, in the region of the pyramidal cell layer of the hippocampus where they both had a peak of activity. For ChAc the maximum activity is in an infra-pyramidal zone, but for GAD we do not have a sensitive enough assay method to decide whether the peak is really in the pyramidal layer or in the infra- or supra-pyramidal zones. However the degeneration studies show that these two enzymes must be localized in different nerve structures.

KERKUT: How sensitive can you get the radioactive assay? Could you measure the ChE and ChAc in a single pyramidal cell?

FONNUM: The maximal sensitivity of the ChAc method is at present 0.25 pmol ACh synthesis/h/μl incubation volume. It is possible to limit this volume down to 0.1 μl and this would make it possible to detect 0.025 pmol/ACh synthesis/h. I do not know if this is enough to detect ChAc activity in a single neurone from mammalian brain. Probably we could determine the activity in a highly active cholinergic neurone.

CROSSLAND: What is the justification for assuming that the presence of GAD in a neurone indicates that the neurone utilizes GABA as an inhibitory transmitter? The concentration of GABA in the brain is high and it does not necessarily function only as an inhibitory transmitter.

Reference p. 58

STORM–MATHISEN: I would like to refer to the Purkinje cells of the cerebellum. The axons of these cells are distributed to the cerebellar nuclei and to the dorsal half of the lateral vestibular (Deiters') nucleus. Here they produce inhibitory postsynaptic potentials that can be mimicked by iontophoretic application of GABA. Further, GABA, but not other amino acids, is released into the 4th ventricle during activation of the Purkinje cells, so it seems that GABA is the transmitter of these cells. We found 2 to 3 times more GAD in the dorsal half of Deiters' nucleus which receives Purkinje terminals than in the ventral half which is devoid of such terminals. We also made lesions of the specific regions in the cerebellar cortex to destroy the Purkinje cells sending their axons to the cerebellar and Deiters' nuclei, and measured the GAD activity. This dropped by a factor of 3 in the locations deprived of their Purkinje terminals, but stayed constant elsewhere. The Purkinje terminals and axons constitute only a small fraction of the tissue, probably no more than 10%. The concentration of GAD in these structures must therefore be enormous (Fonnum, Storm-Mathisen and Walberg, 1970).

FONNUM: Only 30% of the GAD activity in the dorsal part of Deiters' nucleus remains after a lesion in the cerebellar cortex which removes most of the Purkinje cells. It is impossible to destroy all the Purkinje cells so the remaining GAD activity could be ascribed to the Purkinje terminals not destroyed. In Deiters' nucleus, therefore, all the GAD activity is probably found within one type of nerve terminal only, namely those derived from the Purkinje cells. Those terminals, by electrophysiological and pharmacological techniques seem to use GABA as transmitter. This allows us to suggest that GAD is only present in those inhibitory neurones which use GABA as transmitter. There is at present no evidence against making this statement general, namely that GAD is only present in neurones using GABA as a transmitter. Time will show if such a bold statement is correct.

BRADLEY: Did you observe any behavioural effects with your lesions?

STORM–MATHISEN: We did not observe any obvious effects on behaviour, but we did not make tests; and these were unilateral lesions.

REFERENCES

FONNUM, F., STORM–MATHISEN, J. AND WALBERG, F. (1970) Glutamate decarboxylase in inhibitory neurons. A study of the enzyme in Purkinje cell axons and boutons in the cat. *Brain Res.*, **20**, 259–275.

Chemical and Stereochemical Aspects of Behavioural Studies

T. D. INCH

Chemical Defence Establishment, Porton Down, Salisbury, Wiltshire (Great Britain)

Although it is well known that the enantiomers of drugs which possess one or more asymmetric centres often have quite different potencies, the extent of the information that pharmacological studies of enantiomers can provide is often not fully appreciated. Also it is not always appreciated, or at least it is often overlooked, that unless the enantiomeric pair – or better a series of enantiomeric pairs – is carefully chosen and unless the pharmacological test procedures are correctly designed, pharmacological studies of enantiomers will not provide the maximum amount of information. The aim of this paper is first to illustrate the full meaning of these rather general opening remarks by reference to some results we have obtained for some anticholinergic drugs, often referred to as glycollates, and then to suggest that similar studies might be used to clarify certain aspects of the mode of action of other types of drugs which influence behaviour.

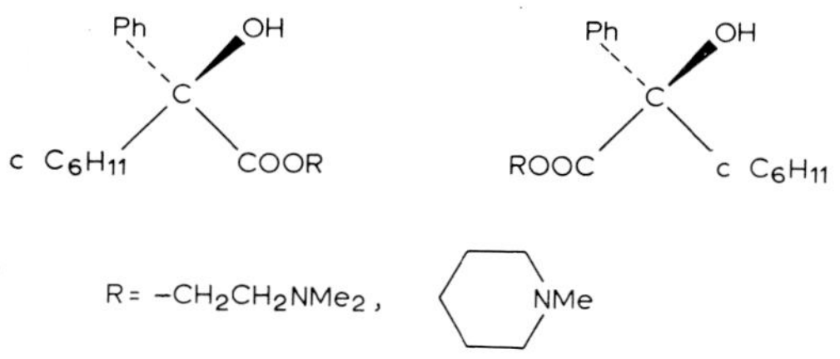

Ph OH C c C_6H_{11} COOR

Ph OH C ROOC c C_6H_{11}

R= $-CH_2CH_2NMe_2$, NMe

Fig. 1. Molecular structure of the dimethylaminoethyl and *N*-methylpiperidin-4-yl 2-cyclohexyl-2-hydroxy 2-phenylacetates.

The compounds we studied are the dimethylaminoethyl and *N*-methylpiperidin-4-yl 2-cyclohexyl-2-hydroxy 2-phenylacetates (Fig. 1).

We prepared the enantiomers and racemates of these compounds both as hydrochlorides and methiodides and so, effectively, we studied the enantiomers and racemate of 4 compounds (Fig. 2).

In order to simplify subsequent interpretation of pharmacological results we went to considerable lengths to obtain the enantiomers of compounds I and II in states of absolute optical purity. That the enantiomers should be optically pure was important because, as will be shown shortly, the differences in the potency of the enantiomers of I and II in some tests was greater than 100 and consequently the presence of as little as 1% of the active R-enantiomers as an impurity in the less active S-enantiomers

References p. 64

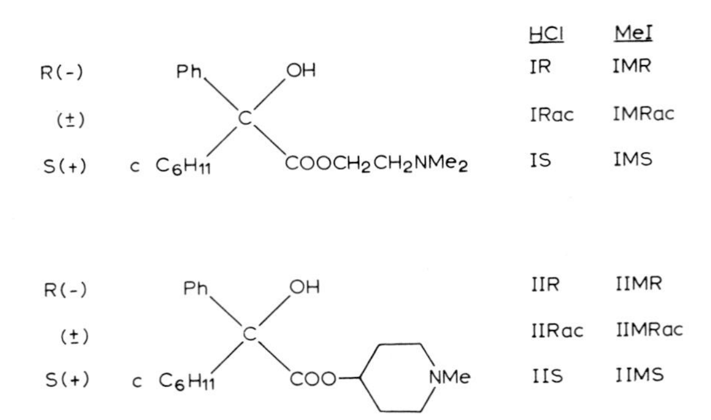

Fig. 2. Molecular structure of the hydrochlorides and methiodides of the enantiomers and racemates of the dimethylaminoethyl *N*-methylpiperidin-4-yl 2-cyclohexyl-2-hydroxy 2-phenylacetates.

could have accounted for most if not all the apparent activity of less the active enantiomer. Since resolution techniques do not necessarily give optically pure enantiomers and since physical methods of analysis are not sufficiently sensitive to detect less than about 2% of optical impurity we devised methods for the stereospecific synthesis of I and II in order to be sure that the enantiomers were optically pure.

We examined the enantiomers and racemates of I and II by 4 tests for anticholinergic potency.

(1) Antagonism of carbachol-induced contractions of guinea pig ileum.
(2) Production of mydriasis in mice.
(3) Antagonism of oxotremorine-induced salivation in mice.
(4) Antagonism of oxotremorine-induced tremor in mice.

(Details of these procedures have been described previously (Brimblecombe *et al.*, 1971).) That is, we used one *in vitro* test, two *in vivo* tests for peripheral anticholinergic potency and one *in vivo* test for central anticholinergic potency.

The results from our studies are summarised in Tables I and II. The results in Table I show that generally the tests were reliable and self-consistent. Thus usually the active R-enantiomers were twice as active as the racemates – with the notable exception of IIMR:IIMRac, the reason for which will be discussed later. Another indication for the reliability of the tests is that quaternization of I caused a similar increase in potency in all the tests. Again compound II showed some deviation from this general trend. In Table II we have shown the R:S enantiomeric potency ratios in all the tests, and we have also given the affinity constants of the racemates. These results show that only for compound I is the potency ratio similar in all the tests and that the *in vivo* potency ratios decrease as the affinity constants increase.

The results from Compound I led to the conclusion that since the R:S potency ratio was the same *in vitro* and *in vivo* and in both central and peripheral potency tests, the anticholinergic receptor must *be the same in the guinea pig ileum, the mouse eye, the mouse salivary gland and in the CNS*. With regard to studies of behavioural effects, the implication of this conclusion is that if the behavioural effects of anticholinergic drugs are directly related to their anticholinergic properties then the ratio

TABLE I

RESULTS OF TESTING THE ENANTIOMERS AND RACEMATES OF COMPOUNDS I AND II FOR ANTICHOLINERGIC POTENCY

Compound	*log k*	*Mydriasis (rel. to atropine)*	*Salivation (μmoles/kg)*	*Tremor (μmoles/kg)*
IR	9.06	0.76	0.76	5.07
I Rac	8.73	0.34	1.40	22.7
IS	7.07	.006	> 100	> 100
IMR	9.66	2.81	0.06	—
IM Rac	9.36	1.67	0.20	—
IMS	7.38	.074	8.84	—
IIR	10.92	2.36	0.18	0.56
II Rac	10.53	1.17	0.32	3.29
IIS	8.48	0.12	7.75	9.42
IIMR	11.08	2.76	0.06	—
IIM Rac	10.39	2.1	0.05	—
IIMS	9.08	1.17	1.05	—

TABLE II

RESULTS OF TESTING THE ENANTIOMERS AND RACEMATES OF COMPOUNDS I AND II FOR ANTICHOLINERGIC POTENCY

Compound	*log k (Rac)*	*log k (ratio)*	*Mydriasis*	*Salivation*	*Tremors*
I	8.73	100	123	> 100	> 22
IM	9.36	100	38	147	—
II	10.53	272	20	43	16.8
IIM	11.08	200	2.3	17.6	

of the behavioural potencies of the enantiomers should approximate to the anticholinergic potency ratio.

At first, the considerable variation between *in vitro* and *in vivo* potencies shown by IM, II and IIM, made us cautious in accepting this implication. However there now seems to be a satisfactory explanation of why enantiomeric potency ratios are different *in vitro* and *in vivo* and why the ratio gets progressively smaller as potency increases. There appears to be a minimum *in vivo* dose which is necessary to produce maximum anticholinergic effects and most of the active R-enantiomers were sufficiently active to produce maximum anticholinergic effects at this low dose. As the potency of the inactive S-enantiomers increases, so the potency ratio becomes smaller.

We think this result emphasizes the point we made right at the beginning when we said it was better to choose a series of enantiomeric pairs for pharmacological comparison rather than just one enantiomeric pair, since conclusions about the stereoselectivity of highly active drugs might be completely misleading. The point we made about the design of test procedures is this: if comparisons of enantiomeric potency

References p. 64

differences are to have any real significance, the test procedures must take into account the time–activity profiles of drug action. It is well known that the rate of action of any drug depends on the dose of drug administered. For enantiomers which differ widely in potency, the rates of action are necessarily different at the ED50 level, and for anticholinergic drugs this difference may be widened still further because even at equimolar doses drugs with high affinity constants require longer to act than drugs with lower affinity constants (Brimblecombe *et al.*, 1971).

Table II shows that *in vivo* potency differences for the mydriasis test were smaller than the potency differences for salivation and tremors. The mydriasis procedure took account of differences in time–activity profiles whereas the salivation test did not. When a time–activity profile for the salivation test was determined, the IIR : IIS ratio was close to that observed in the mydriasis test. (It is possible that some of the variation in the enantiomeric potency ratios in the mydriasis and salivation tests arose because of differences in the method of administration of the drugs in the two tests, *viz.* intravenous injection in the mydriasis experiments and intraperitoneal injection in the salivation and tremor experiments). Differences in the time–activity profiles of IR and IS were not critical, since within the time scales of most tests both enantiomers were fast acting drugs.

Having recognised many of the factors responsible for the observed enantiomeric potency ratio variations we are now confident that there is little variation in the anticholinergic receptor at the different sites assayed by the 4 procedures we have used.

Behavioural studies of IIR and IIS have indicated that the R-enantiomer is about 10 times more active than the S-enantiomer. This potency ratio is close to that observed *in vivo* for the anticholinergic ratio and thus if we can demonstrate a similar parallelism with other compounds, and particularly with IR : IS, strong evidence will have been provided that the behavioural effects of glycollates are directly related to their anticholinergic properties.

In the above discussion we have tried to illustrate that there are considerable difficulties associated with the acquisition of significant data from enantiomeric comparisons, but we believe that the results it is possible to obtain fully justify the effort involved. In addition to providing information about the similarity or otherwise of receptors, enantiomeric comparisons also provide a critical test of certain concepts of the chemical nature of drug action.

For example, it has been suggested (Abood, 1970) that glycollates may act by a carbonium ion mechanism (Fig. 3). The facts that the enantiomers act at different rates and have different durations of action rule out this idea since during the formation of the carbonium ion the asymmetry of the enantiomer is destroyed.

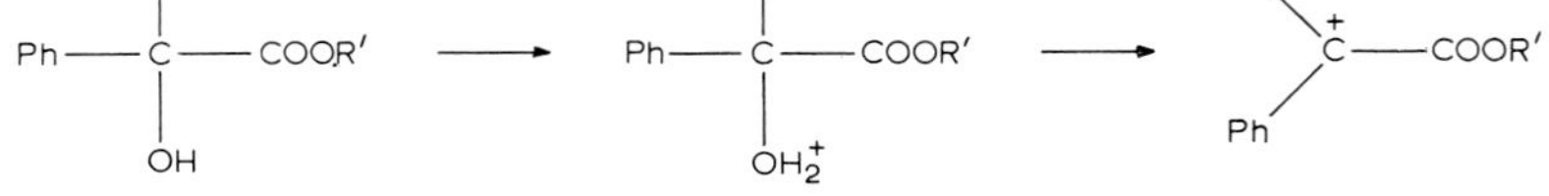

$$\mathrm{Ph{-}C(R)(OH){-}COOR'} \longrightarrow \mathrm{Ph{-}C(R)(OH_2^+){-}COOR'} \longrightarrow \mathrm{(R)(Ph)\overset{+}{C}{-}COOR'}$$

Fig. 3. The action of glycollates by a carbonium ion mechanism.

Our studies have also invalidated the hypothesis that the lone pair of nitrogen electrons in glycollate bases are essential to anticholinergic and psychotomimetic activity (Abood, 1970; Gabel and Abood, 1965), since quaternary salts were more active than the hydrochlorides and there was no evidence that central effects were caused by a significantly different mechanism from peripheral effects.

To move away from studies of anticholinergic drugs we would first like to remind you that the value of enantiomeric comparisons has been demonstrated by Taylor and Snyder (1970) with studies of (+)- and (−)-amphetamine. These workers showed that (+)-amphetamine was markedly more potent as an inhibitor of catecholamine uptake by noradrenaline neurones in the brain than was (−)-amphetamine, whereas the two isomers were equally active in inhibiting catecholamine uptake by the dopamine neurones of the corpus striatum. In behavioural studies (+)-amphetamine was 10 times as potent as (−)-amphetamine in enhancing locomotor activity while it was only twice as potent in eliciting a compulsive gnawing syndrome. This suggested that the locomotor stimulation induced by amphetamine involved central noradrenaline, whereas dopamine neurones played an important role in the induced compulsive gnawing behaviour.

The experiments that have been described for the comparisons and (+)- and (−)-amphetamine probably only scratch the surface of the problem but they certainly provide an indication that from similar studies much more information could be obtained to differentiate behaviours involving noradrenaline and dopamine in the brain.

We now want to give one example of a potential use of enantiomeric comparisons for establishing whether or not different types of drug produce behavioural effects in similar ways. There has been considerable discussion about whether or not drugs such as (+)-lysergic acid diethylamide (LSD), certain N,N-dimethyltryptamines and mescaline (and other methoxylated amphetamines) have the same kind of CNS activity. Some authors have suggested that these drugs do act in a similar fashion and considerable attention has been focussed on the stereochemical and electronic similarities of these drugs (Fig. 4) (Snyder and Richelson, 1968; Chothia and Pauling, 1969). Viewed in this way it is clear that they could interact with a common receptor.

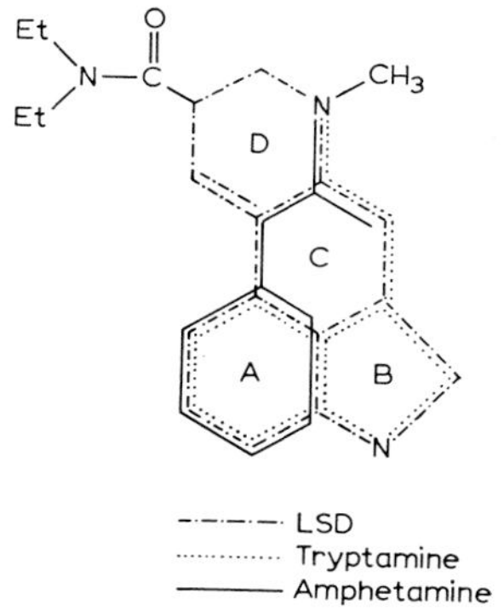

Fig. 4. Congruence of the molecular structure of LSD, tryptamine, and amphetamine.

References p. 64

Evidence in favour of this concept could be provided by a comparison of enantiomeric potency ratios. If there was good agreement between the potency ratios of, for example (+)- and (−)-LSD, (+)- and (−)-α-methyl-tryptamine and (+)- and (−)-amphetamine, and if the fundamental molecular geometry of the (+)-isomers was similar, the contention that these drugs act by a similar mechanism would be strengthened.

ACKNOWLEDGEMENTS

The work described in this paper was carried out in collaboration with Dr. R. W. Brimblecombe, Mr. D. M. Green and Mr. D. A. Buxton. Technical assistance was provided by Mrs. P. B. J. Thompson and Miss R. A. L. Powers.

REFERENCES

ABOOD, L. G. (1970) Stereochemical and Membrane Studies with Psychotomimetic Glycolate Esters. In D. H. EFRON (Ed.), *Psychotomimetic Drugs*, Raven Press, New York, pp. 67–74.

BRIMBLECOMBE, R. W., GREEN, D. M., INCH, T. D. AND THOMPSON, P. B. J. (1971) The significance of differences in the potency of enantiomers of antiacetylcholine drugs, *J. Pharm. Pharmacol.*, **23**, 745–757.

CHOTHIA, C. AND PAULING, P. (1969) On the conformations of hallucinogenic molecules and their correlation. *Proc. nat. Acad. Sci. (Wash.)*, **63** ,1063–1070.

SNYDER, S. H. AND RICHELSON, E. (1968) Psychedelic drugs: steric factors that predict psychotropic activity. *Proc. nat. Acad. Sci. (Wash.)*, **60**, 206–213.

TAYLOR, K. M. AND SNYDER, S. H. (1970) Amphetamine: differentiation by D- and L-isomers of behaviour involving norepinephrine or dopamine. *Science*, **168**, 1487–1489.

DISCUSSION

COHEN: The study of the effects of stereo-isomers is of great importance to pharmacology. You were quite right in stressing the fact that pure enantiomers should be used. Another important point is that one should be sure that no racemization *in vivo* occurs. Working with sarin we had difficulty in avoiding this pitfall since racemization of this compound is catalysed by fluoride ions which arise from the hydrolysis of sarin by sarinase in rat blood plasma.

INCH: Yes, this is true. Fortunately, in the case of the glycollates they are fairly stable molecules anyhow and one can, I think, check that racemization is not occurring by careful comparison of the R enantiomer to the racemate in this case. As long as a 2:1 ratio is obtained, racemization is unlikely to have occurred.

BRADLEY: Do you think you are justified in suggesting that the central and peripheral receptors for ACh are similar on the basis of potency studies? The test used for central effects was one for muscarinic activity only. Surely there could be qualitative as well as quantitative differences *vis à vis* the nicotinic and muscarinic nature of the receptors.

INCH: Yes, but I think that this is just one more piece of information to go with all the rest on this matter. It should not be viewed in isolation by any means.

BRIMBLECOMBE: It is probably worth saying that we obtained similar time profiles with these drugs both *in vivo* and *in vitro*, *i.e.*, those drugs which took a long time to reach equilibrium in the guinea-pig ileum experiments for the determination of log k_b values were also those which took a long time to reach peak effects in the mydriasis experiments.

Changes in the Properties of Acetylcholinesterase in the Invertebrate Central Nervous System

G. A. KERKUT, P. C. EMSON, R. W. BRIMBLECOMBE, P. BEESLEY, G. W. OLIVER AND R. J. WALKER

Department of Physiology and Biochemistry, Southampton University, Southampton, and C.D.E., Porton Down, Salisbury, Wiltshire (Great Britain)

In an insect such as the cockroach there are two main chemical transmitters in the CNS. Acetylcholine (ACh) is the main excitatory transmitter and excites over 95% of all the cells tested and γ-aminobutyric acid (GABA) is the main inhibitory transmitter and inhibits over 95% of all the cells tested (Colhoun, 1963; Treherne, 1966; Boistel, 1968; Kerkut *et al.*, 1969; Pitman and Kerkut, 1970; Pitman, 1971). Although there are other transmitters present in the insect ganglion such as dopamine and noradrenaline (Hodgson and Wright, 1963; Klemm, 1968; Frontali, 1968; Frontali and Haggendal, 1969; Mancini and Frontali, 1970), ACh and GABA appear to be the dominant transmitters and the system in the insect is so much more simple than that present in the vertebrate's spinal cord where there are possibly 7 different transmitters, or in the gastropod nervous system where there are at least 5 different transmitters. Thus the insect CNS is pharmacologically more simple than that of the vertebrates or the molluscs.

We decided to study changes in the chemistry of the insect nervous system following behavioural changes. We used the preparation devised by Horridge (1962) where he set up two cockroaches so that one got a shock every time its leg touched

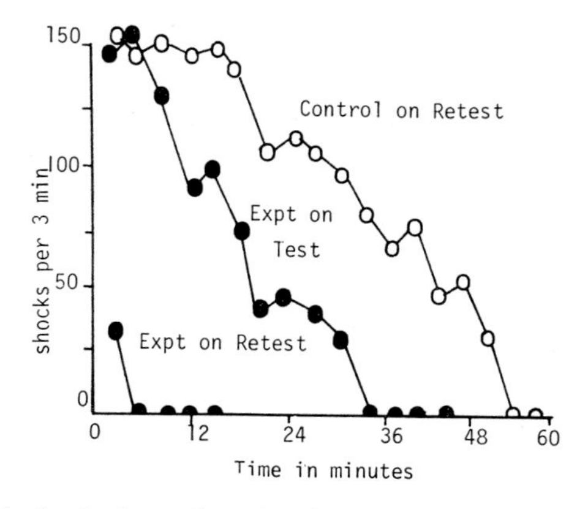

Fig. 1. Avoidance conditioning in the cockroach. The Experimental animal learned within 36 min to keep its leg out of the solution. The Control animal required 50 min on retest to learn to keep its leg out of the solution.

References pp. 76–77

the saline (the Experimental animal), and a second cockroach (the Control) was yoked to the first so that it got a shock every time the first cockroach got a shock, regardless of the position of the leg. The first cockroach learned to keep its leg out of the saline whereas the second animal, even though it got the same number of shocks as the first, did not learn to keep its leg out of the saline.

We (Kerkut *et al.*, 1970) investigated this preparation and found the following results. When insects were set up and trained in this way, the Experimental animal learned to keep its leg out of the solution within about 36 min and then on retest showed that it required only a very few shocks in order to keep its leg out of the solution. The Control on retest, however, required a considerable number of shocks before it learned to keep its leg out of the solution. In Fig. 1 the animal required over 50 min trial before it learned to keep its leg out of the solution. Although both groups of animals received the same number of shocks on initial training, the Experimental animal learnt to keep its leg out of the solution and the Control did not.

Effect of drug injection

Some drugs, such as cycloheximide, if injected into the cockroaches 1 h before training, slowed up the training procedure and made the animal require many more shocks (Fig. 2). Other drugs, such as neostigmine, if injected into the animal 1 h

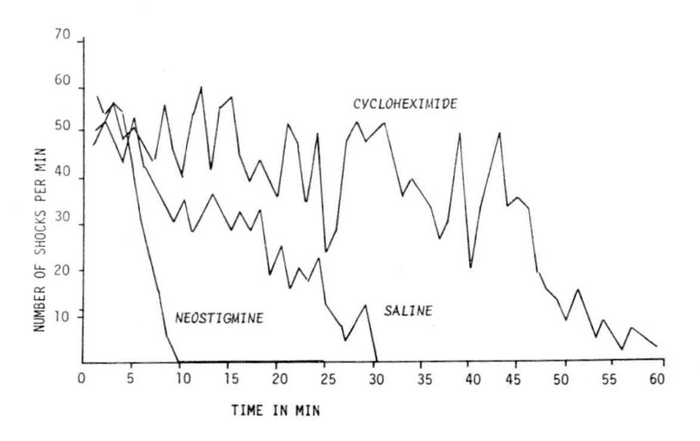

Fig. 2. The effect of cycloheximide and neostigmine on learning. The drugs were injected into the animal 1 h before testing. Neostigmine allowed the animal to learn more quickly than the animals injected with saline.

before training speeded up the acquisition of learning by the Experimental animal so that it could learn within 10 min instead of the normal 33 min that the saline-injected Experimental animal required. The drugs that facilitated learning such as prostigmine, edrophonium, pemoline or amphetamine all showed a dose–response relationship. Fig. 3 shows the amount of drug injected into the animal 1 h before training and the numbers of shocks that it received before it had reached the learning criterion of keeping its leg out of the solution. The normal Experimental animal required 300 shocks, but with an injection of 150 μg of prostigmine it could learn after 50 shocks. The effects of these drugs give some indication as to possible bio-

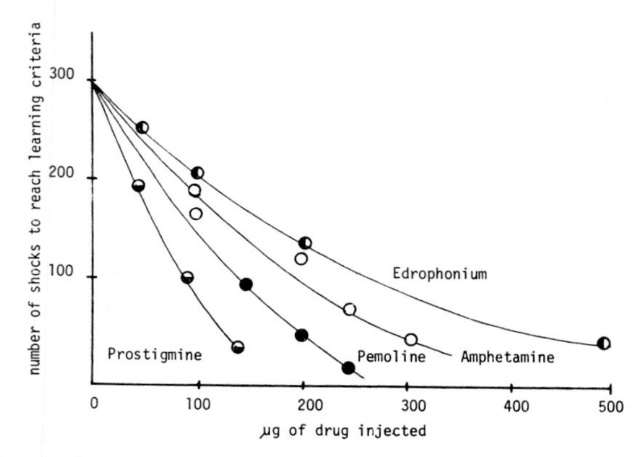

Fig. 3. The effect of drug dose on the number of shocks required for the cockroach to reach learning criteria. The normal Experimental animal learnt after about 300 shocks. Prostigmine allowed the animal to learn after only 30 shocks. The effect was dose dependent.

chemical intermediates in the learning process. Pemoline is related in structure to the nucleotides and possibly could affect the synthesis of ribonucleic acid (RNA). Amphetamine affects the catecholamines in the CNS and could possibly work via transmitters such as dopamine and noradrenaline. Prostigmine and edrophonium are anticholinesterases. We decided first to investigate the cholinergic system in the insect ganglion.

Cholinesterase activity in the insect ganglion

When we measured the cholinesterase (ChE) activity in the metathoracic ganglia of the cockroach we found that the Resting ganglia, *i.e.*, the ganglia from the animal that was pinned down but not stimulated, had a level of 3.42 µmoles of acetic acid released/h/100 µg protein. The Control ganglia which had received shocks but had not learnt had a level of 2.43, while the Experimental ganglia had a level of 1.22 (Table I).

TABLE I

THE ACTIVITY OF AChE IN THE METATHORACIC GANGLIA OF THE COCKROACH. (AChE EXPRESSED IN TERMS OF µMOLES ACETIC ACID RELEASED/h/100 µg PROTEIN)

Treatment group	*AChE activity*
Experimental	1.22 ± 0.08 (n = 21)
Control	2.43 ± 0.19 (n = 21)
Resting	3.42 ± 0.53 (n = 21)

Thus the level of ChE in the Experimental animal was much lower than that of the Control or Resting animal. The decrease was greater than 50%. There was also a change in the ChE activity in the head of the cockroach even though the metathoracic segment had been trained. Table II (Oliver *et al.*, 1971) shows that the ChE activity

TABLE II

THE ACTIVITY OF AChE IN THE HEAD OF THE COCKROACH AFTER THE METATHORACIC LEG HAS BEEN TRAINED

Treatment group	*AChE activity*
Experimental	2.30 ± 0.11 (n = 5)
Control	2.81 ± 0.08 (n = 5)
Resting	3.00 ± 0.10 (n = 10)

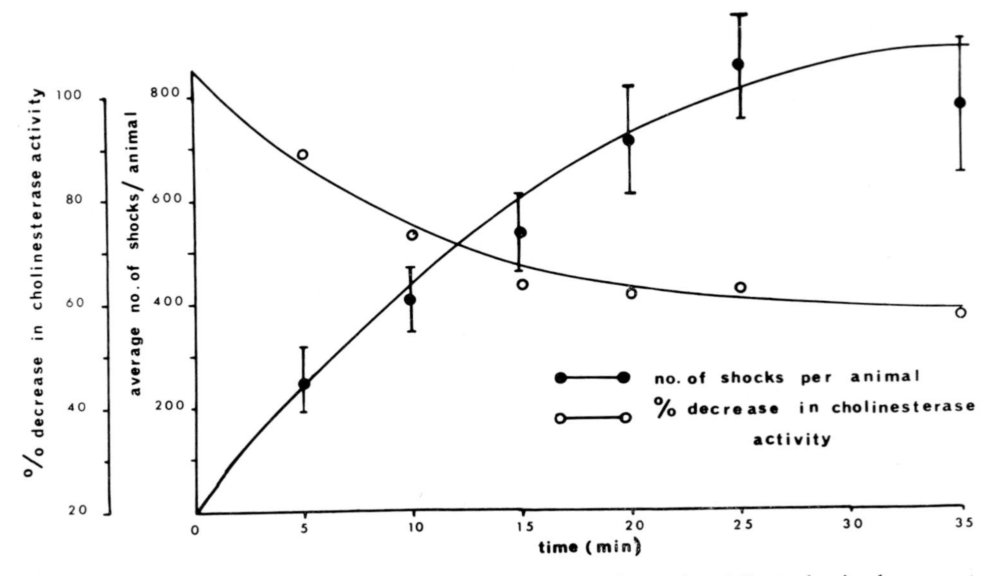

Fig. 4. The change in ChE activity during learning. The Experimental and Control animals were set up for a given period of time (5, 10, 15 etc., min). The number of shocks that they received is plotted, together with the difference between the ChE activity in the Experimental ganglia. As the animals received the shocks, there was a greater decrease in the activity of the ChE in the Experimental animals.

in the head of Resting cockroaches had a level of 3, that of Control cockroaches had a level of 2.81 and that of the Experimental cockroaches, *i.e.*, those that had learned, had a level of 2.30. The esterase activity in the head of the Experimental animal was lower than that of the Control or Resting animals. Fig. 4 shows the change in the ChE concentration in the metathoracic ganglion whilst the animal is being trained. A series of animals was trained for times ranging from 5, 10, 15, 20 etc., to 35 min. At each 5-min interval, a group of animals was taken and the ChE activity of the metathoracic ganglion was measured. These values were compared to the level of a Control group trained for a similar time, and the curve shows the difference between the Experimental and the Control, the Control for each time being taken as 100%. There was a steady fall in the ChE concentration of the Experimental ganglia over 35 min. The second line shows the number of shocks that the preparations received

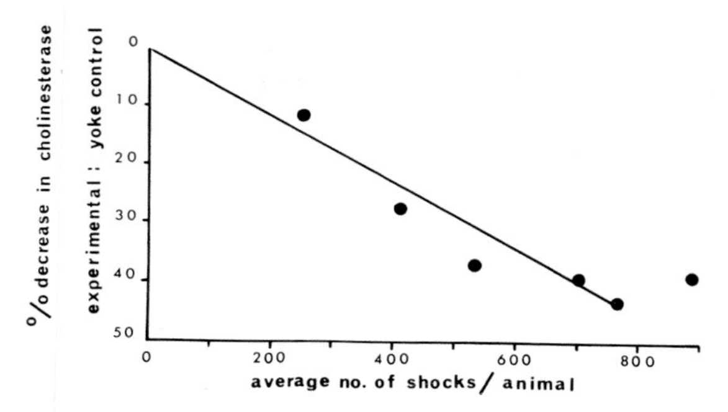

Fig. 5. Relation between ChE activity in the Experimental animals and the number of shocks received.

and as the animals learned, *i.e.*, they received fewer shocks with time, so the esterase level flattened off. The esterase level fell 40% in 20 min. There was a slight lag phase of about 2 min before the esterase dropped. Fig. 5 shows the relationship between the number of shocks that the preparation received and the percent decrease in the ChE level of the Experimental *vs* yoked Control animal. As the number of shocks increased, so the activity of the esterase decreased. Again there was an initial lag phase before there was any change in the ChE concentration. Thus as the Experimental animal learnt over the 20–30 min there was a rapid fall in the ChE activity in the ganglia.

Forgetting

If a group of cockroaches was trained and then tested the next day it was found they required a few shocks in order to retrain them back to learning criteria. If they were left for 2 days they required more shocks, and if they were left for 3 days they returned to the initial naive state. Such a training procedure is shown in Fig. 6. The animal was trained on day 0 and then retested on days 1, 2 and 3. The number of minutes required for retraining is shown on the right ordinate. Over the 3 days the animals gradually forgot the training procedure. A similar group of animals were analysed

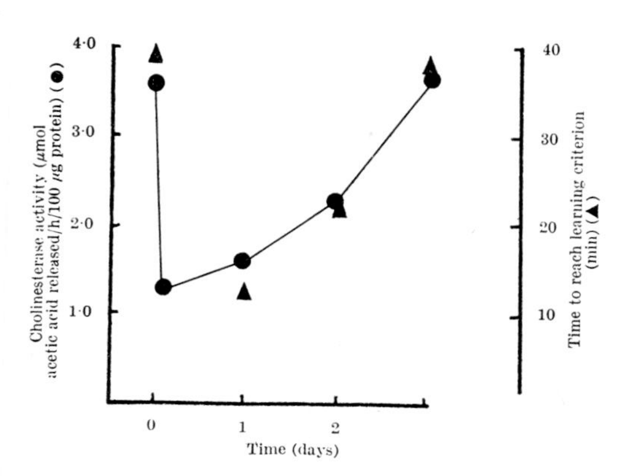

Fig. 6. Relationship between the ChE in the Experimental animals (●) and their forgetting (▲). As the ChE activity increased, the animals forgot.

References pp. 76–77

for the ChE activity of the metathoracic ganglia. These are shown in circles on Fig. 6. On training, the ChE activity dropped and then over the 3 days it slowly recovered to the initial level. There was a very good correlation between the recovery of the esterase activity and the forgetting of the animal.

The biological significance of the changing ChE activity during the training procedure is discussed in more detail in another paper (Kerkut *et al.*, 1970). We should like here to consider some features of the nature of the ChE change itself.

Recovery of cholinesterase activity

Fig. 7 shows a photograph of a polyacrylamide separation of proteins in the cockroach ganglia homogenate. It is possible to cut the polyacrylamide gel into slices and to measure the ChE activity in these slices. Such a curve is shown in Fig. 8. There are two regions of high AChE activity. The activity in the homogenate from the Control ganglia is shown by the dashed lines and there is a high activity in the slow

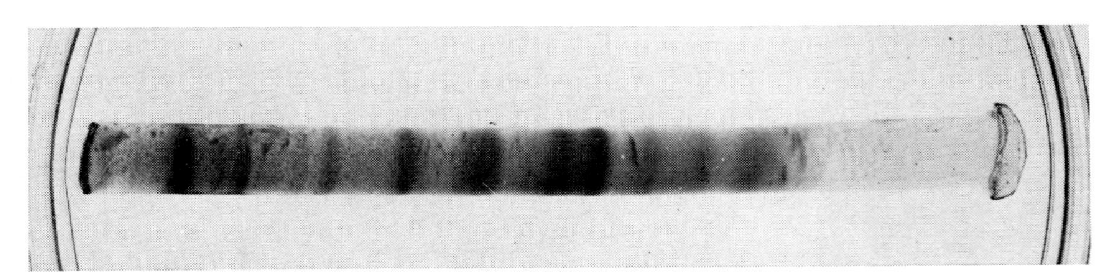

Fig. 7. Polyacrylamide separation of the proteins in the homogenates of the cockroach thoracic ganglia.

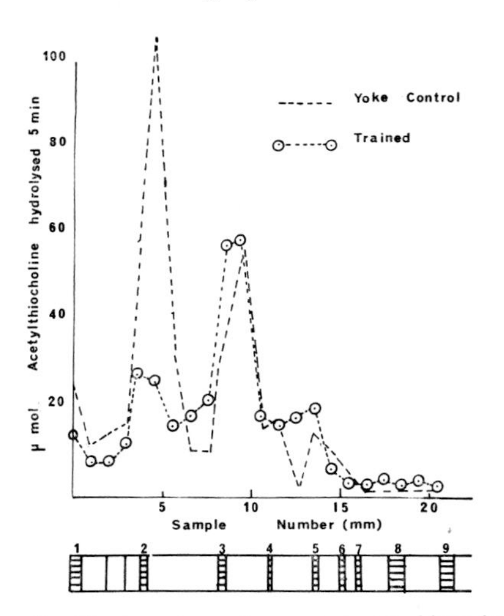

Fig. 8. ChE activity in the ganglia. The homogenate was separated into its protein bands by polyacrylamide electrophoresis. The ChE activity relative to the protein concentration is plotted. Two bands of ChE activity were found (bands 2 and 4). There was a difference in the ChE activity in band 2 between that of the Experimental animals' ganglia (○–○–○) and that of the Control (– – –). The ChE of the Experimental animals had less activity.

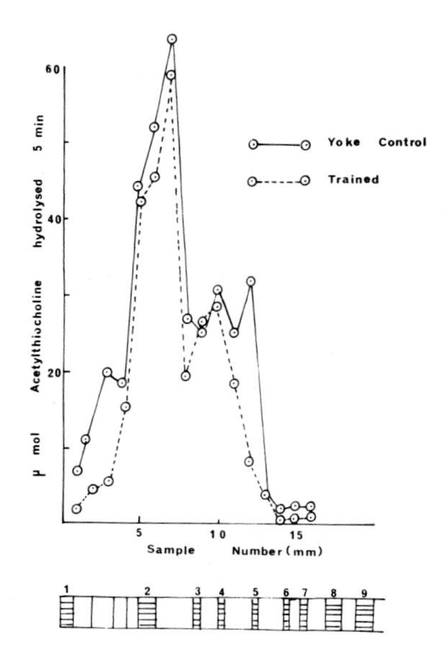

Fig. 9. ChE activity in the ganglia. The gel was identical to that shown in Fig. 8 except that it was left for 48 h in phosphate buffer before the ChE activity was determined. The Experimental ganglia in band 2 than had the same activity as the Control.

fraction and a moderate activity in the faster fraction of the ChE. The activity in the homogenate of the Experimental ganglia is plotted on the same scale and there is a decrease in the activity in the slow fraction whilst the activity of the fast fraction is the same. This means that the difference between the Experimental and the Control ganglia is mainly shown by one of the ChE fractions. If the polyacrylamide gel is left for 48 h before we estimate the ChE activity we get a different result (Fig. 9). Now the slow fraction ChE activity, derived from the Experimental ganglia, has risen and equals that of the Control ganglia. Note, however, that the maximal level is 60 instead of 100. This result agrees with that obtained from the dialysis of the whole homogenate, but we have the additional information that it is the slow fraction ChE that has changed its activity in the Experimental ganglia.

Cholinesterase levels vary from species to species and from strain to strain within a species, and also with the age of the animal (Augustinsson, 1963; Usdin, 1970). Isoenzymes have been demonstrated for pseudocholinesterase on numerous occasions (Webb, 1964). The reports of isoenzymes among acetylcholinesterases are fewer. Ecobichon and Israel (1967) found 4. Isoenzymes of acetylcholinesterase in the electric eel. Isoenzymes of acetylcholinesterase seem common in insects. They have been demonstrated in the cricket (Edwards and Gomez, 1966), house fly and cockroach (Eldefrawi *et al.*, 1970) and the mayfly (Krysan and Kruckeberg, 1970).

The nature of the change in the cholinesterase

There are 5 possible reasons for the decreased ChE activity in the cockroach ganglia. These are:

(1) Reduced synthesis of ChE.
(2) Substrate inhibition.
(3) End product inhibition.
(4) Production of an anticholinesterase.
(5) Ionic changes leading to conformational change.

We are still investigating the nature of the change in the ChE but one indication is given by a study of the kinetics of the enzyme activity. Fig. 10 shows a Lineweaver

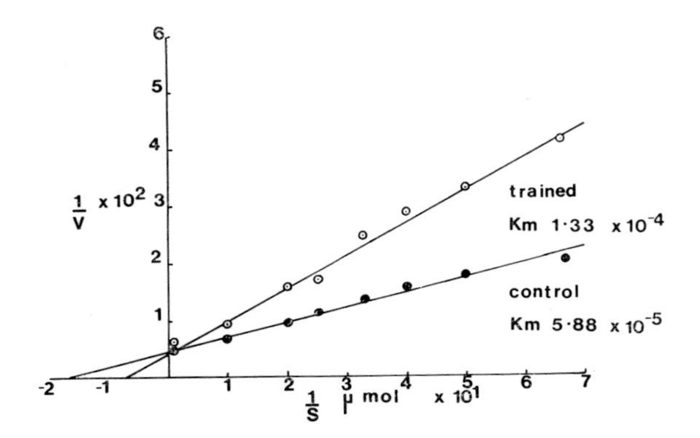

Fig. 10. Lineweaver Burk plot of activity of ChE in Experimental (○–○–○) and Control (●-●-●) cockroaches.

Burk plot of the activity of the enzyme homogenates of the Experimental ganglia and the Control ganglia. The two enzymatic preparations had the same V_{max} but they had different K_ms. The K_m for the Experimental ganglia was 1.33×10^{-4} moles whilst that for the Control was 5.88×10^{-5} moles. This suggests that the difference between the enzyme in the two preparations is that probably both have the same number of active sites (same V_{max}) but that there is a difference in the affinity between the active site and the substrate.

Fluorescence of acridine orange

Another indication of the difference between the ChE in the Experimental and Control ganglia is shown by use of acridine orange. In these experiments a homogenate of the ganglia was mixed with a fixed amount of acridine orange. Fig. 11a shows the fluorescence of the acridine orange when it was mixed with the Control ganglia homogenate or the Experimental ganglia homogenate. The fluorescence that is shown was mainly proportional to the amount of free acridine orange in solution. The higher peak on the Experimental ganglia indicates that less acridine orange was bound to the enzyme in the Experimental ganglia. The nature of the binding of the acridine orange to the esterase is indicated in Fig. 11b where acridine orange and neostigmine were added.

If the acridine orange bound onto the enzyme, the fluorescence was reduced. The fluorescence present when neostigmine was added first was very much greater than if the acridine orange was added first. This suggests that there may be some compe-

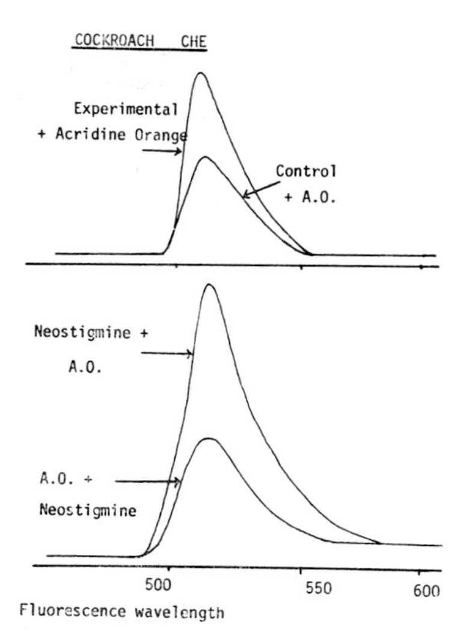

Fig. 11. The effect of ChE from Experimental and Control animals on the binding of acridine orange. The ChE from the Experimental animals bound less acridine orange than did the ChE from the Control animals. There is competition for binding between acridine orange and neostigmine onto the ChE. A.O. acridine orange.

tition between neostigmine and acridine orange for the active site of the enzyme and therefore the site where the acridine orange acts could be the same as the neostigmine site: in other words they compete for the ACh site. This experiment, however, needs considerable refinement before the results can be taken as being truly quantitative.

Learning in snails

We carried out a similar series of experiments to those just described in the cockroach for snails *Helix aspersa*. The training set-up is shown in Fig. 12. In this case when we measured the ChE activity in the brains of the Experimental, Control and Resting snails we found a striking difference from that of the cockroaches. In the snail there

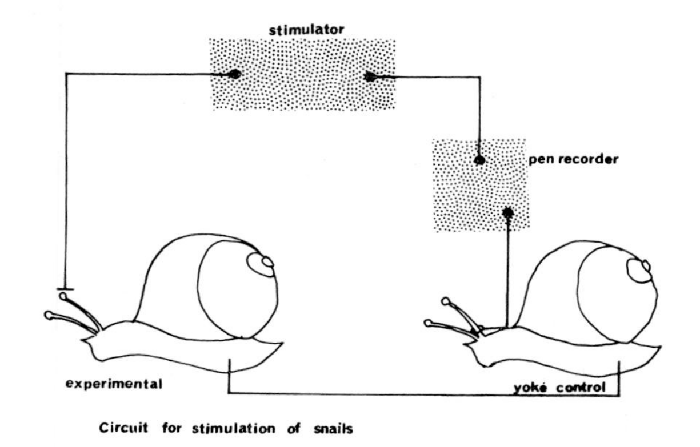

Fig. 12. Set-up for training snails in avoidance conditioning.

References pp. 76–77

was an increase in the activity of ChE in the ganglia of the Experimental animals, the increase being 50% greater in the Experimental than in the Control and 100% greater in the Experimental than in the Resting animal (Table III). When the snails forgot, which they did over 24 h, there was a change in the ChE activity (Fig. 13). On training the esterase activity rose and then over the 24 h, as the animals forgot, the esterase activity fell back to the original level. Fig. 14 shows a Lineweaver Burk

TABLE III

THE ACTIVITY OF AChE IN THE BRAIN OF THE SNAIL

Treatment group	*AChE activity*
Experimental	0.592 ± 0.009 (n = 7)
Control	0.407 ± 0.027 (n = 7)
Resting	0.245 ± 0.023 (n = 7)

In the snail brain ACh is mainly an inhibitory transmitter. In the cockroach, however, ACh is mainly excitatory.

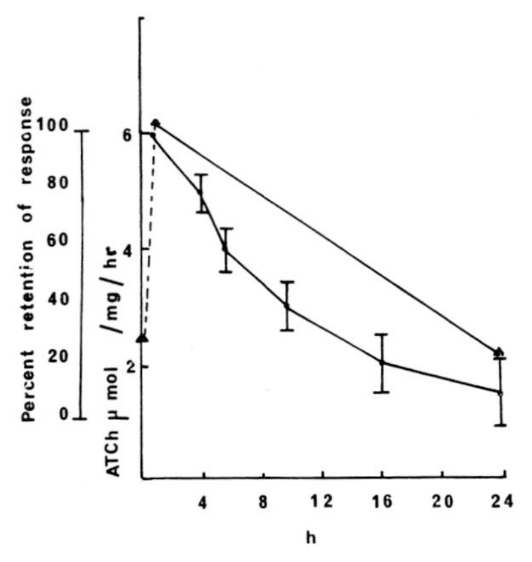

Fig. 13. ChE and forgetting in snails. When snails learnt, there was an increase in the activity of their ChE (▲). As they forgot there was a decrease in the activity of the ChE back to normal levels (-●-●-).

plot of the activity of the snail enzyme in the brains of Experimental animals and Control animals. Here the Experimental animals had the greater activity and also a different K_m from that of the Control. The Control K_m was 8.44×10^{-4} mole, whilst that of the Experimental was 5.66×10^{-4} mole. Both enzymes had the same V_{max}. This suggests that again there is the same number of active sites in the enzyme of the Experimental and Control but that there is a difference in the affinity for the substrate so that the Experimental enzyme has the greater affinity for the substrate.

By means of sodium dodecyl sulphate polyacrylamide separation it is possible to split the ChE in the snail brain down into 4 separate units (Fig. 15). These have molecular weights of 37,800, 46,000, 70,000 and 92,000, giving an aggregate total molecular weight for the whole enzyme of approximately 200,000. Molecular weight estimations

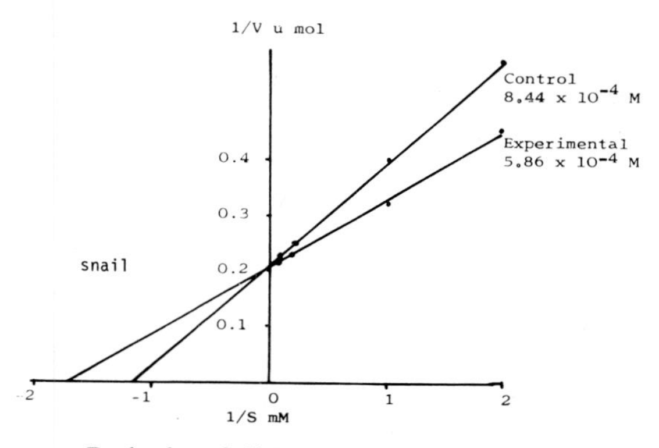

Fig. 14. Lineweaver Burk plot of ChE activity in Experimental and Control animals.

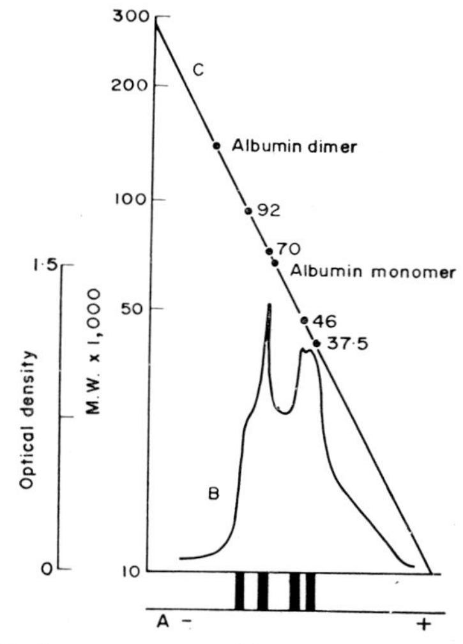

Fig. 15. Separation of the ChE after treatment with SDS, on polyacrylamide gel. A, the pattern of the ChE activity in the gel. There are 4 bands of ChE activity. B, The plot of the optical activity of these bands. C, The distance travelled by the proteins in the gel. An albumin monomer and dimer is included. The molecular weights of the 4 ChE bands are given.

have been done on the purified enzyme from electric eel. Leuzinger *et al.* (1969) found that the enzyme consists of two identical sub-units. Each sub-unit consists of two polypeptide chains, an alpha and beta chain. Krysan and Kruckeberg (1970) have estimated the molecular weight of the house fly, honey bee and mayfly cholinesterase as 160,000 molecular weight units. This was on the basis of data from sucrose gradient centrifugation. They observed that the cholinesterase would aggregate to give higher molecular weight polymers. This is a well known feature of acetylcholinesterase and depends on the ionic strength of the medium (Grafius and Millar, 1967).

The function of a synapse is to stop transmission and normally there are many presynaptic impulses before one postsynaptic impulse is set up. Thus if there are 20 presynaptic action potentials setting up one postsynaptic action potential, the synaptic ratio would be 20:1. In the cockroach, if the ChE decreased, the efficacy of the

References pp. 76–77

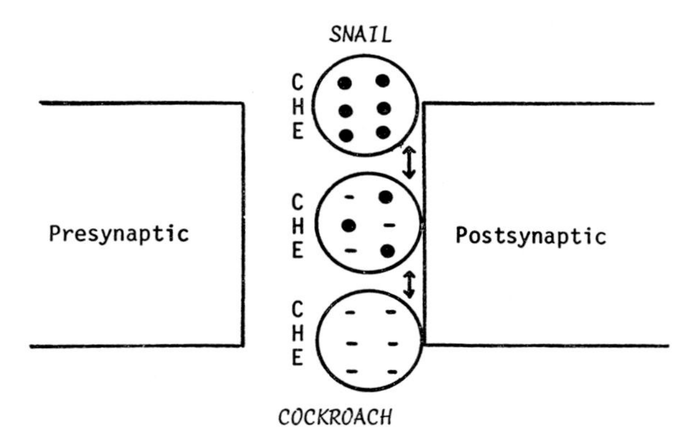

Fig. 16. Diagrammatic representation of the change in ChE activity in the synapse. It is suggested that the number of active sites remains the same but that the activity can change so as to increase (●), as in the snail, or decrease (—), as in the cockroach.

ACh would be greater and the synaptic ratio would possibly change from 20:1 to 5:1 or even 1:1, thus bringing about a facilitatory pathway. In the snail, where ACh is mainly an inhibitory transmitter, the increased activity of the enzyme would change the synaptic ratio from 20:1 to 40:1, *i.e.*, it would decrease the amount of inhibition and so bring about greater facilitation. A diagrammatic representation of this phenomena is shown in Fig. 16.

These results are of interest because they show the way in which a given enzyme in the nervous system could effectively control the longer-term, *i.e.*, 3 min to 3 days, changes in the activity of the CNS. We would not suggest that they are by any means a complete answer because it is quite clear that there are changes in other transmitter systems occurring at the same time. There are also changes in enzyme levels and protein synthesis occurring in the nervous system following patterned stimulation. The elucidation of these changes will give greater understanding of the nature of the changes that take place during short-term and long-term memory of the CNS.

ACKNOWLEDGEMENTS

We are indebted to the Wellcome Trust and to the Ministry of Defence for grants supporting some of this work.

REFERENCES

AUGUSTINSSON, K.-B. (1963) Classification and comparative enzymology of cholinesterases and methods for their determination. *Hand. exp. Pharmacol.*, **15**, 89–128.

BOISTEL, J. (1968) The synaptic transmission and related phenomena in insects. *Advanc. Insect. Physiol.*, **5**, 1–64.

COLHOUN, E. H. (1963) The physiological significance of acetylcholine in insects and observations upon other pharmacologically active substances. *Advanc. Insect Physiol.*, **1**, 1–46.

ECOBICHON, D. J. AND ISRAEL, Y. (1967) Characterisation of the esterases from the electric tissues of *Electrophorus* by starch gel electrophoresis. *Canad. J. Biochem.*, **45**, 1099–1105.

EDWARDS, J. E. AND GOMEZ, D. (1966) Bound cholinesterase in the nervous system of *Acheta domesticus*. *J. Insect Physiol.*, **12**, 1061–1068.
ELDEFRAWI, M. E., TRIPATHI, R. K. AND O'BRIEN, R. D. (1970) Acetylcholinesterase isoenzymes from the housefly brain. *Biochim. biophys. Acta (Amst.)*, **212**, 308–314.
FRONTALI, N. (1968) Histochemical localisation of catecholamine containing neurones in the cockroach brain. *J. Insect Physiol.*, **14**, 881–886.
FRONTALI, N. AND HAGGENDAL, J. (1969) Noradrenaline and dopamine content in the brain of the cockroach *Periplaneta americana*. *Brain Res.*, **14**, 540–542.
GRAFIUS, M. A. AND MILLAR, D. B. (1967) Reversible aggregation of acetylcholinesterase. II. Interdependence of pH and ionic strength. *Biochemistry*, **6**, 1034–1046.
HODGSON, E. S. AND WRIGHT, A. M. (1963) Action of epinephrine and related compounds upon the insect nervous system. *Gen. comp. Endocr.*, **3**, 519–525.
HORRIDGE, G. A. (1962) Learning of leg position by the ventral nerve cord of headless insects. *Proc. roy. Soc. B*, **157**, 33–52.
KERKUT, G. A., PITMAN, R. M. AND WALKER, R. J. (1969) Iontophoretic application of ACh and GABA onto insect central neurones. *Comp. Biochem. Physiol.*, **31**, 611–633.
KERKUT, G. A., OLIVER, G. W. O., RICK, J. T. AND WALKER, R. J. (1970) The effects of drugs on learning in a simple preparation. *Comp. gen. Pharmacol.*, **1**, 437–483.
KLEMM, N. (1968) Monoaminehaltige Strukturen im Zentralnervensystem der Trichopteren (Insecta) Teil I. *Z. Zellforsch.*, **92**, 487–502.
KRYSAN, J. L. AND KRUCKEBERG, W. C. (1970) The sedimentation properties of the cholinesterase from a mayfly *Hexagenia bilineta*, Ephemeroptera and the Honeybee *Apis mellifera*. *Int. J. Biochem.*, **1**, 241–247.
LEUZINGER, W. M., GOLDBERG, M. AND CAUVIN, E. (1969) Molecular properties of acetylcholinesterase. *J. molec. Biol.*, **40**, 217–225.
MANCINI, G. AND FRONTALI, N. (1970) On the ultrastructural localisation of catecholamines in the Beta lobes (Corpora pedunculata) of *Periplaneta americana*. *Z. Zellforsch.*, **103**, 341–350.
OLIVER, G. W. O., TABERNER, P. V., RICK, J. T. AND KERKUT, G. A. (1971) Changes in GABA level, GAD and ChE activity in CNS of an insect during learning. *Comp. Biochem. Physiol.*, **38**, 529–535.
PITMAN, R. M. (1971) The pharmacology of insects. *Comp. gen. Pharmacol.*, **2**, 347–371.
PITMAN, R. M. AND KERKUT, G. A. (1970) Comparison of the actions of iontophoretically applied acetylcholine and gamma aminobutyric acid with the EPSP and IPSP in cockroach central neurones. *Comp. gen. Pharmacol.*, **1**, 221–231.
TREHERNE, J. E. (1966) *The Neurochemistry of Arthropods*. Cambridge University Press.
USDIN, E. (1970) Reactions of cholinesterase with substrate inhibitors and reactivators. In *International Encyclopedia of Pharmacology and Therapeutics*, Section 13, Pergamon Press, Oxford, pp. 49–354.
WEBB, E. C. (1964) The nomenclature of multiple enzyme forms. *Experientia (Basel)*, **20**, 592–593.

DISCUSSION

BRADLEY: Your idea that a change in the amount of ChE reflects facilitation of the neurone is dependent upon the amount of ACh released remaing constant; it is possible that the changes in ChE are simply related to more or less ACh being released presynaptically, *i.e.*, the more ACh released, the more ChE will be required to destroy it.

KERKUT: This is true, but on the other hand it is possible to record on the postsynaptic side, and we find by comparing the Experimental with the Control that there is very much more activity in the Experimental cockroach which is coming through on the postsynaptic side. I think, however, that we must measure levels of ACh and ChAc.

ANSELL: What surprises me is that postulated subtle conformational changes in the AChE are retained through the excitation procedure and enzyme assay.

KERKUT: The surprising thing in the cockroach enzyme, or even in that of the snail, is how stable the system is. Oxyhaemoglobin or haemoglobin are fairly stable providing you keep the conditions right. I think it is an oversimplification to assume that everything always comes apart during extraction.

Reference p. 78

FONNUM: Does not the effect of cycloheximide indicate that synthesis of AChE is of importance? Is AChE membrane bound as in mammals or is it soluble? Did you detect any difference in localization of the different isoenzymes? Did you measure any change in the ACh level?

KERKUT: Sixty percent of the AChE is membrane bound, but we think that it is the soluble enzyme which changes. We cannot explain the effect of the cyclohexamide in terms of AChE, and we do not think it necessarily relates to AChE synthesis because the change in the snail is too quick. There is no alteration in the polyacrylamide localization of AChE when Experimental and Control animals are compared. We are running a series of experiments at present to measure ACh.

STORM–MATHISEN: Is this AChE pre- or postsynaptic? In the CNS of mammals, AChE seems to be presynaptic according to the data from electromicroscopy denervation experiments and subcellular fractionation studies.

KERKUT: Electron microscopy localization has not been precise but it seems that the enzyme is both pre- and postsynaptic (Newman *et al.*, 1968).

HEILBRONN: If it can be assumed that your method of treating tissue prior to AChE-determination does not change the amounts of free or membrane bound AChE present, it may be possible to tell if the ratio between the two changes in your learning experiments. It is still a matter of discussion whether the K_m values of free and membrane bound AChE differ. Personally, I think there is evidence that they do. The two K_m values you observed could therefore represent values for free and membrane bound AChE, respectively.

KERKUT: Thank you. That is a very interesting suggestion.

FONNUM: We (Laake, Andersen and Fonnum, unpublished observations) have recently studied the inhibition of AChE from frog, rat and chicken brain. To our surprise, the inhibition rate showed large variation in that chicken AChE was inhibited 300 times faster than frog AChE with dyflos and 10 000 times faster with a quaternary organo-phosphate compound. It therefore seems that the inhibition rate of an organo-phosphorous compound is very sensitive to small changes in AChE molecules and may be used to detect conformational changes in the AChE molecules which you have described.

WATTS: Is it necessary to dialyze AChE from the trained cockroaches in order to get the increase in activity with incubation in buffer solution?

KERKUT: No, it is not necessary. If the separated polyacrylamide gel is left in buffer for 48 h, much of the activity returns to the specific band.

WATTS: Is the AChE from the trained snails deactivated under the same conditions?

KERKUT: We have not as yet studied the mechanism of the loss of activity in the snail enzyme. The behavioural studies linked with AChE measurements indicate that activity returns to resting level after 24 h.

REFERENCES

NEWMAN, KERKUT, G. A. AND WALKER, R. J. (1968) *Symp. zool. Soc. Lond.*, **22**, 1–17.

Hallucinogenic Drugs and Circadian Rhythms

J. A. DAVIES, R. J. ANCILL AND P. H. REDFERN

Pharmacology Group, School of Pharmacy, The University of Bath, Bath, Somerset (Great Britain)

In common with all living matter, the brain is subject to rhythmic changes in its functional state. The most common, and most marked rhythms are the "24-h rhythms", so called because they are synchronised by the 24 h light–dark cycle of the environment (for examples of such rhythms see Richter, 1964; Mills, 1966; Friedman and Walker, 1968). Many of these rhythms can be shown to exist solely as a result of environmental factors, and disappear in the absence of any external time cue (*Zeitgeber*). Others are endogenous rhythms, which under normal circumstances are conditioned to an exact 24 h cycle by external *Zeitgebers*. However, in the absence of time cues, these rhythms do not disappear, but revert to their true frequency, which only approximates to 24 h. These are the true *circadian* rhythms.

The variation over 24 h in the concentration of biochemical constituents of brain tissue, and in particular of possible neurotransmitter substances is now well established, (Friedman and Walker, 1968; Scheving *et al.*, 1968; Dixit and Buckley, 1967) and it would therefore seem reasonable to expect that the response to centrally acting drugs might also be subject to a significant variation with clock-hour. Surprisingly little attention has been paid to this aspect of central pharmacology; the few published reports have been concerned with drugs which affect the level of consciousness, such as fluothane (Mathews *et al.*, 1964), pentobarbitone (Pauly and Scheving, 1964) and Lidocaine (Lutsch and Morris, 1967), and there is virtually no information available on the effects of psychotropic drugs.

Our initial intention was therefore to investigate the extent to which responses to psychotropic drugs were affected by the time of administration. We chose a group of "psychotomimetic" drugs because it seemed possible that the effects of such agents might be related to circadian rhythms from an additional standpoint: it is well known that many forms of mental illness are accompanied by disturbances of the normal sleep/wakefulness cycle (Conroy *et al.*, 1968), and it is assumed that such disturbances are secondary manifestations of the disease. However, it has also been amply demonstrated that in mentally normal individuals, temporary disruption of external *Zeitgebers* may lead to abnormal behaviour, as evidenced by the effects of transatlantic flight (Proceedings of the Syntex Symposium 1970). If, by analogy, it is postulated that disturbances of circadian rhythms are involved in the aetiology of psychosis then the mode of action of the "psychotomimetic" drugs may be related to their well-known ability to produce time distortion (Fischer, 1967). We have therefore tried to answer two questions:

References pp. 94–95

(1) To what extent are the effects of psychotomimetic drugs affected by the clock-hour at which they are administered?

(2) Are psychotomimetic drugs capable of affecting circadian rhythms?

CLOCK-HOUR AND THE RESPONSE TO PSYCHOTOMIMETIC DRUGS

In this first part of the investigation, we looked at the effect of a group of "sympathomimetic psychotomimetic" drugs (Bebbington and Brimblecombe, 1969), mescaline, lysergic acid diethylamide (LSD) and amphetamine on locomotor activity, a parameter known to be subject to a marked 24-h rhythm.

Methods

Male Sprague–Dawley rats weighing 175–200 g were housed in a soundproof chamber for at least 10 days before each experiment. The animals were subjected to a 12-h light–12-h dark cycle, with the light period from 06.00 to 18.00 h. The temperature in the chamber was 22 ± 1 °C. The animals were fed and watered daily at a randomly chosen clock-hour during the light phase, access to food and water being *ad libitum*.

On the day of the experiment, locomotor activity was recorded from different groups of rats for 2-h periods beginning at 19.00, 23.00, 03.00, 07.00, 11.00, 15.00 and 19.00 h. Animals were injected intraperitoneally immediately before the start of each

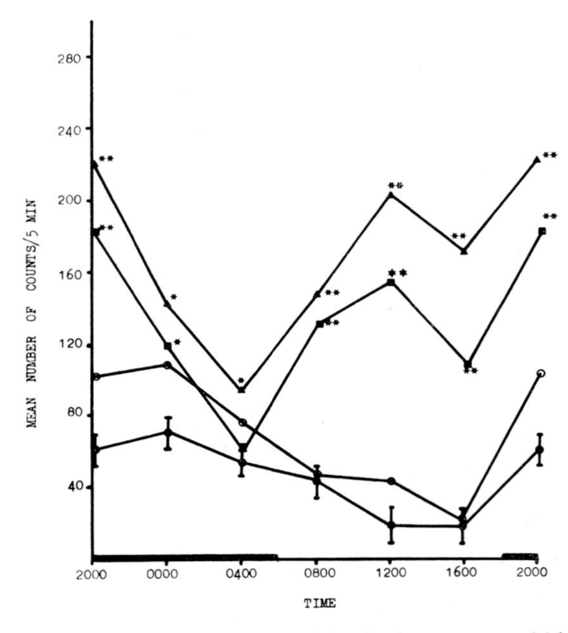

Fig. 1. The effect of amphetamine on locomotor activity in the rat over a 24-h period. (●——●) control, 0.9% saline; (■——■) 1.25 mg/kg; (▲——▲) 2.5 mg/kg; (○——○) 10 mg/kg. * significant at $P = 0.05$ level; ** significant at $P = 0.01$ level.

period either with the test drug or with normal saline, the injection volume being 0.2 ml. The animals were then placed in activity cages in which the floor was made up of 30 evenly spaced stainless steel bars, insulated from each other and being alternately earthed and live. The bridges that the animal made or broke with its paws were converted into pulses and summed, emerging as an automatic print-out every 5 min. The activity recorded was therefore chiefly locomotor, but any action which involved lifting the paws from the floor, *e.g.*, washing, rearing, scratching, also contributed to the final score. Each cage was enclosed in a sound-proof cabinet which excluded the print-out mechanism and which was lit on the same cycle as the environmental chamber. In the first experiments (those involving mescaline), groups of 3 animals were placed in each cage; for all subsequent experiments, activity was recorded from individual animals. The form of the drugs used was mescaline hydrochloride (Sigma), (±)-amphetamine hydrochloride (Sigma) and lysergic acid diethylamide (Brocades, Great Britain). All doses were calculated in terms of base.

Amphetamine. The effect of amphetamine on locomotor activity is shown in Fig. 1.

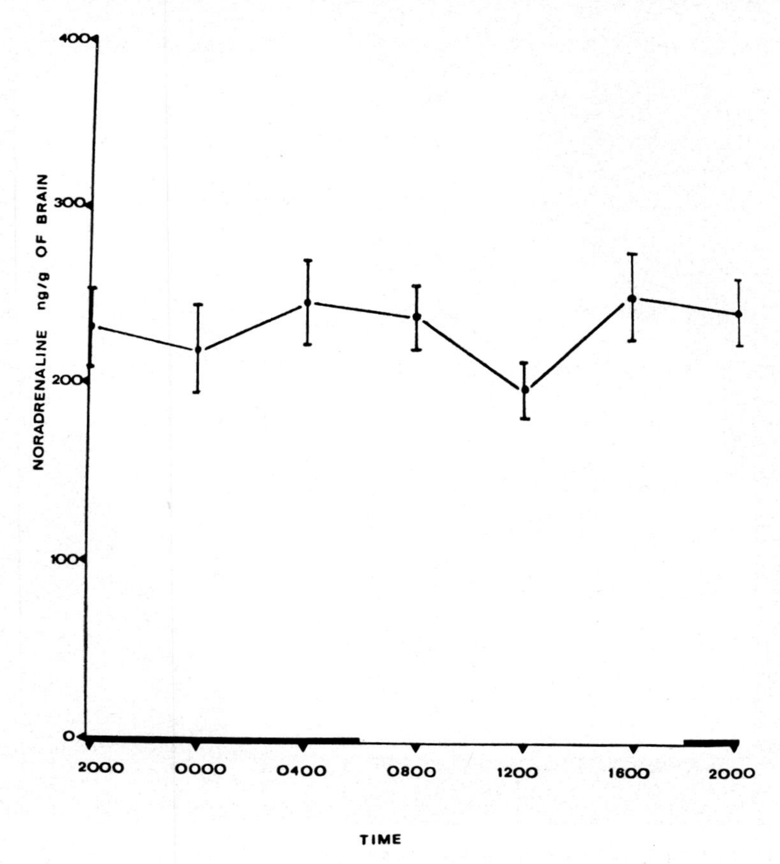

Fig. 2. Variation with time in the noradrenaline concentration (ng/g) of rat brain. Each point represents the mean from 30 animals.

References pp. 94–95

The recorder used printed out a "total activity count" every 5 min. The first 4 of these, representing initial exploratory behaviour, were discarded, and a "mean activity count" was then obtained for the remaining 100 min. Statistical analysis was carried out using the Two-sample rank test (Goldstein, 1964) and Student's *t*-test.

The control group exhibited the expected 24-h rhythm, being more active during the dark phase than during the light phase. Similarly, the response to amphetamine during the light phase was expected (Del Rio and Fuentes, 1969). It could be seen, however, that the response to amphetamine varied significantly with clock-hour. The variation had an ultradian pattern with a trough at the end of the dark phase (04.00 h) and a second trough, less marked but nonetheless significant, 12 h later at 16.00 h. It may be relevant to point out that the concentration of noradrenaline in the brain shows a similar ultradian rhythm, as can be seen in Fig. 2. This curve represents the mean of 7 separate experiments, performed at each clock-hour, and the resulting rhythm has a small amplitude. However, when the results from each experiment are plotted separately (Fig. 3), the amplitude in each experiment is seen to be much larger.

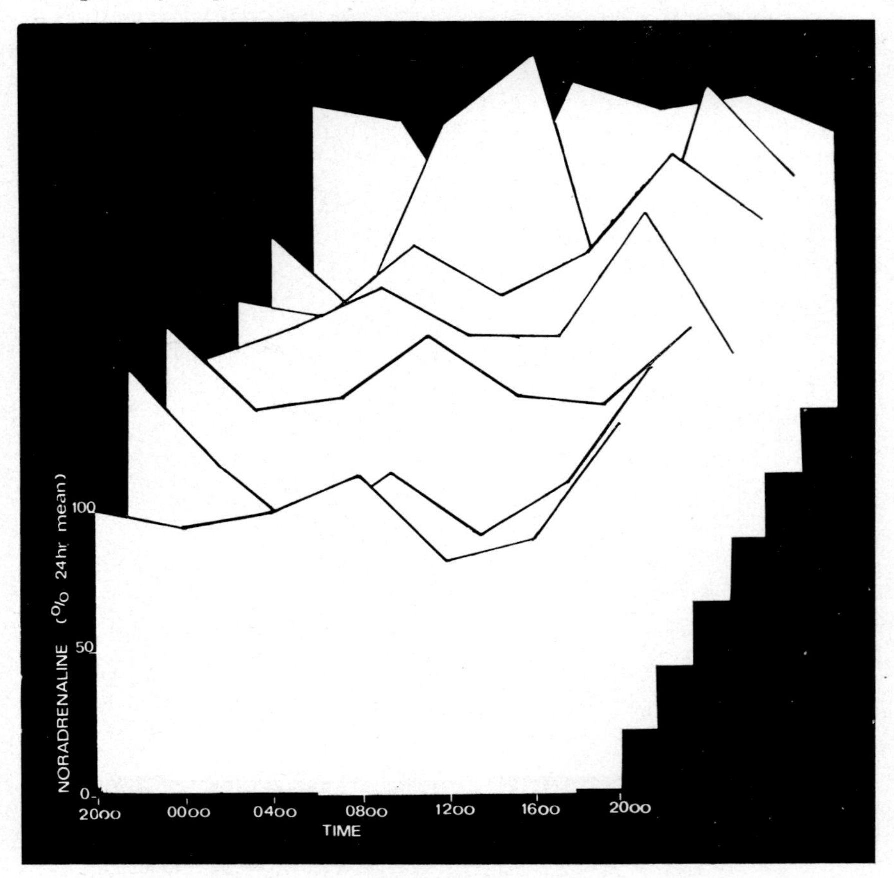

Fig. 3. Variation with time in the noradrenaline concentration of rat brain. The results from 7 experiments have been separately represented. In each experiment, 5 animals were used at each clock-hour.

Small shifts in the time scale between experiments mean that the summed effect is less impressive. We have no information as to why brain noradrenaline concentration varies with clock-hour. Since the form of the rhythm is different from that of locomotor activity, it presumably cannot be simply related to the sleep-wakefulness cycle. This problem is at present under investigation in our laboratory.

LSD. The effect of LSD on locomotor activity is shown in Fig. 4. The expected increase in locomotor activity during the light phase (Cohen and Wakely, 1968) occurred. This increase in locomotor activity was greatest at the beginning of the dark phase, but for the rest of the dark phase LSD produced no significant response. Thus

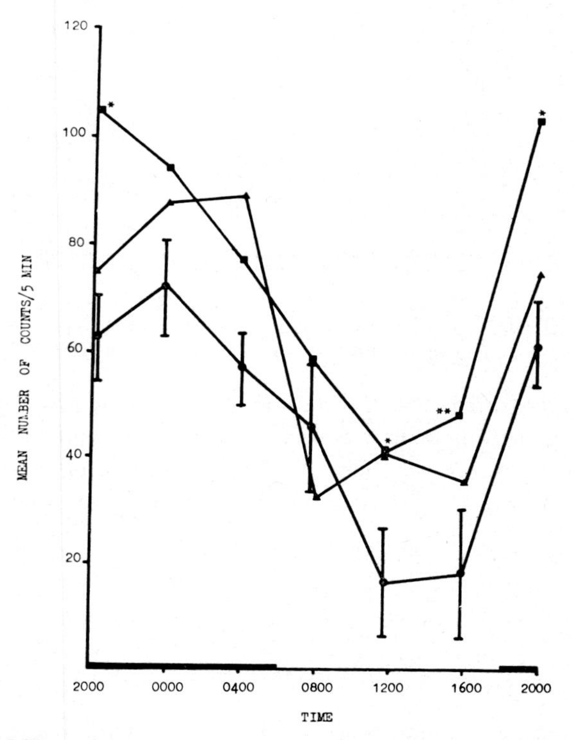

Fig. 4. The effect of LSD on locomotor activity in the rat over a 24-h period. (●———●) control, 0.9% saline; (■———■) 100 pg/kg; (▲———▲) 200 pg/kg.

the response to LSD was greatest during the period when the animals are normally least active, and when brain 5-hydroxytryptamine levels are highest (Scheving *et al.*, 1968; Dixit and Buckley, 1967; Ancill *et al.*, 1971).

Mescaline. Unlike LSD, mescaline produced a greater response during the dark phase, as shown in Fig. 5. Only during this phase did 12.5 mg/kg mescaline i.p. produce a significant increase in locomotor activity.

Thus, we have demonstrated a marked variation with clock-hour in the response to this group of psychotomimetic drugs. Perhaps it is not surprising, given the nature

References pp. 94–95

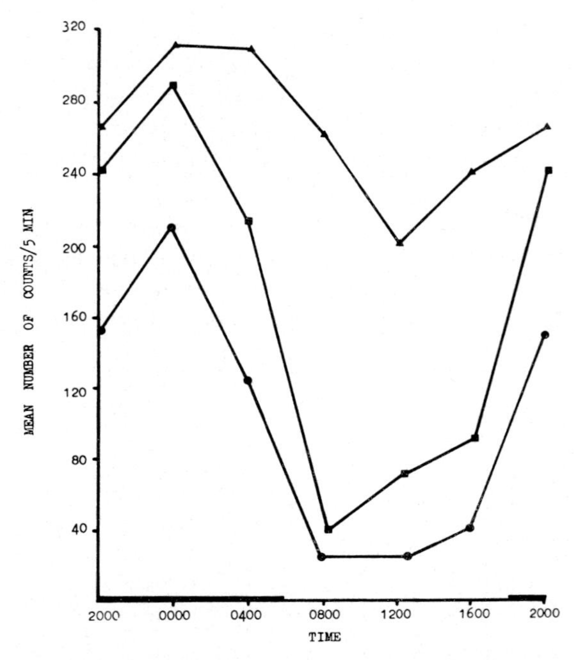

Fig. 5. The effect of mescaline on locomotor activity in the rat over a 24-h period. (●———●) control, 0.9% saline; (■———■) 12.5 mg/kg; (▲———▲) 25 mg/kg.

of the parameter measured, that no common pattern emerged. Nevertheless, from these preliminary experiments there is some indication that the variation in response is associated with endogenous rhythms. The significance of this association is difficult to assess since the causes of the circadian variation in brain amine concentrations are not known.

THE EFFECT OF PSYCHOTOMIMETICS ON 24-H RHYTHMS

Methods

In these experiments a continuous record of locomotor activity was obtained from animals caged individually in the soundproof, temperature-controlled chamber described above. Each rat was kept in a "Squirrel-Cage" (E. K. Bowman Ltd., London) which consisted of an activity wheel, 16 in. in diameter and with a 5 in. rim, attached to a cage measuring 8 in. × 10 in.

There was continuous access between the wheel and the cage. A revolution of the wheel in either direction closed a microswitch, thus activating a small induction coil and thereby causing a pen, attached to a metal level held adjacent to the coil, to make a mark on the recording chart. To avoid feedback, the recorder was kept outside the environmental chamber. Feeding and watering were carried out every third day at a random clock hour, so that disturbance was minimal, and provided no regular *Zeitgeber*.

At the start of each series of experiments, activity was recorded under these control conditions until a regular pattern of activity was established. The time taken for this to occur depended on the illumination cycle being used. The water supply was then replaced by the drug solution for 8–14 days. Finally recording was continued, after removal of the drug solution and the reintroduction of water, until the original rhythm returned. Thus each animal acted as its own control. At each dose level at least 4, and usually 6, animals were used.

Light–dark cycle

Fig. 6 shows a typical response to an amphetamine solution of 0.125 mg/ml, in rats subjected to a 12-h light/12-h dark cycle. From the amount of solution consumed by each animal it is possible to calculate that the mean dose was 3 mg/kg/day, although

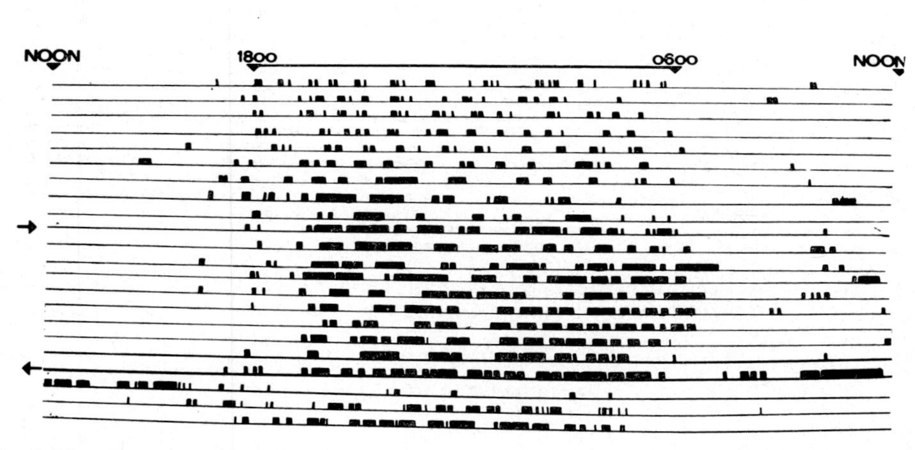

Fig. 6. The effect of amphetamine (0.125 mg/ml) on the 24-h rhythm of locomotor activity in the rat. lighting cycle: 12 h/light/12 h dark.(→) Introduction of drug solution;(←) removal of drug solution.

we have no means of knowing whether there was an even consumption throughout the period of administration. It can be seen that during the first 9 days under control conditions the normal 24-h rhythm of locomotor activity was apparent, with movement almost entirely restricted to the dark phase. On the introduction of amphetamine there was an increase in activity, but again the increase was restricted to the dark phase. The activity rapidly returned to normal when the amphetamine solution was replaced by water after 10 days.

A 0.25 mg/ml solution of amphetamine (6 mg/kg/day) produced an increase in activity during both the light and the dark periods (Fig. 7). It is interesting to note that activity during drug administration was not continuous throughout 24 h; the two periods during which activity was consistently absent (Noon–16.00 h and 0.600–08.00 h) are approximately the periods when the acute effects of amphetamine were least (Fig. 1) and when central noradrenaline levels were highest (Fig. 2). This effect was rapidly reversed when the amphetamine solution was removed. An even higher dose of amphetamine, 0.5 mg/ml (20 mg/kg/day), produced an increase in activity throughout 24 h, so that the 24-h rhythm disappeared, but reappeared when the drug solution was removed (Fig. 8).

In contrast, neither LSD 5 μg/ml (230 μg/kg/day) nor mescaline 1.25 mg/ml (50 mg/kg/day) had any effect on the light-synchronised 24-h rhythm (Figs. 9 and 10).

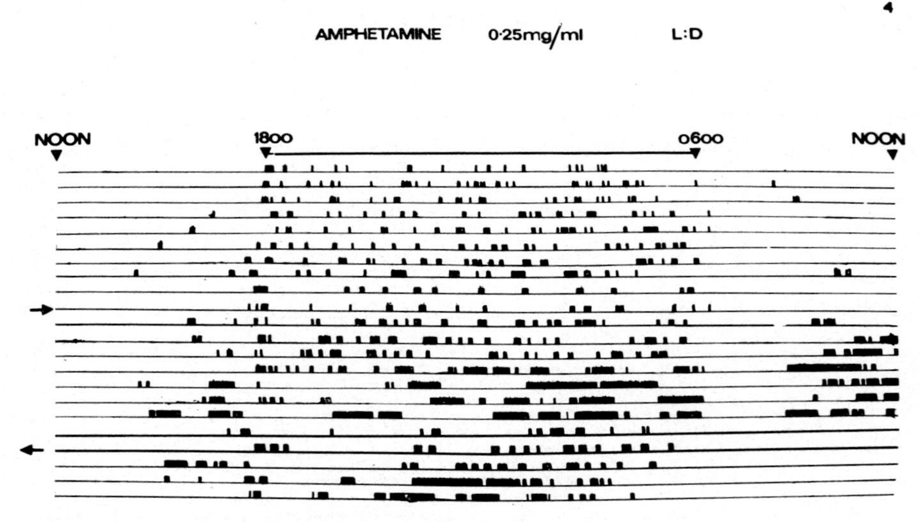

Fig. 7. The effect of amphetamine (0.25 mg/ml) on the 24-h rhythm of locomotor activity in the rat. Lighting cycle: 12 h light/12 h dark.(→) Introduction of drug solution;(←) removal of drug solution.

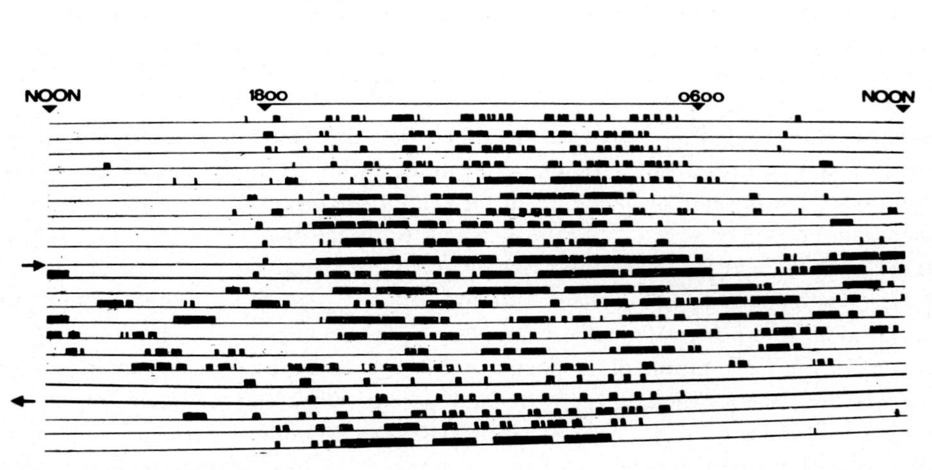

Fig. 8. The effect of amphetamine (0.5 mg/ml) on the 24-h rhythm of locomotor activity in the rat. Lighting cycle: 12 h light/12 h dark.(→) Introduction of drug solution;(←) removal of drug solution.

Continuous light

When animals are kept in an environment which is devoid of all time cues, the endogenous circadian rhythms which are inherent in and specific to each individual, become exposed. Quay (1968) showed that under such conditions animals kept in full sight and hearing of each other were not synchronised by each other, but exhibited individual rhythms. A similar observation has been made in humans (Kleitman, 1963).

Fig. 11 is the record obtained from an animal kept in continuous light. It can be seen that during the control period of 21 days, the circadian rhythm of the animal, freed

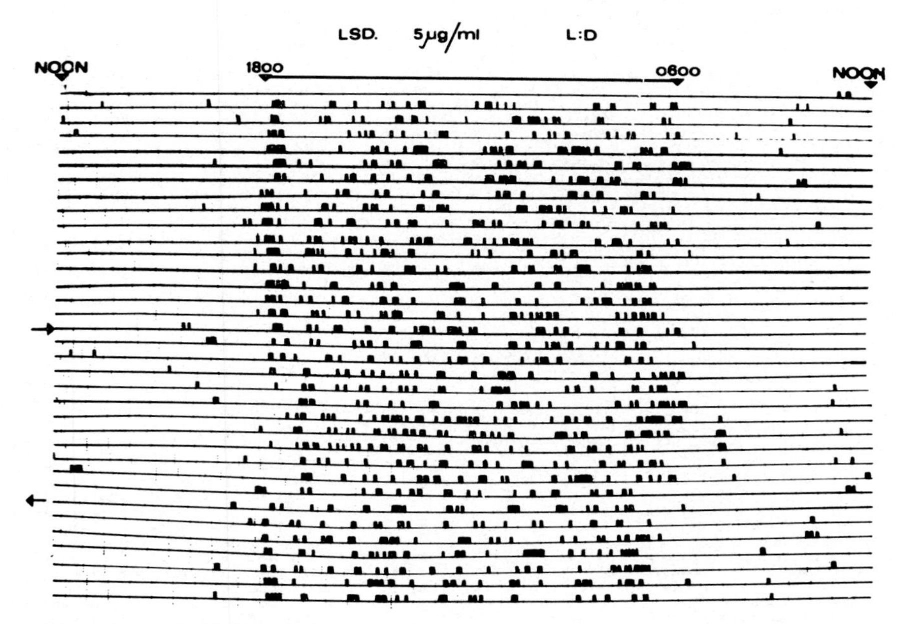

Fig. 9. The effect of LSD (5 μg/ml) on the 24-h rhythm of locomotor activity in the rat. Lighting cycle: 12 h light/12 h dark.(→) Introduction of drug solution;(←) removal of drug solution.

from all synchronisers, assumed its inherent length, represented by a shift to the right in the activity period when measured against a 24-h time scale. As expected, the rhythms exhibited in continuous light were all longer than 24 h (Aschoff, 1963), having lengths of between 24 h 58 min and 25 h 53 min. It seemed reasonable to assume, in view of this variation, that all *Zeitgebers* had been successfully excluded. The introduction of amphetamine (0.125 mg/ml) between the 21st and 35th days produced an increase in activity within the confines of the normal activity period only. Both 0.25 mg/ml amphetamine solution (Fig. 12) and 0.5 mg/ml amphetamine solution (Figs. 13 and 14) caused complete disruption of the circadian rhythm of activity, an effect which disappeared on withdrawal of the drug solution. Comparison of Fig. 12 with Fig. 7 shows that the effect of 0.25 mg/ml amphetamine was more marked in animals under conditions of continuous light, indicating that the circadian rhythm is more easily disrupted than the 24-h rhythm. It is also interesting to compare Figs. 13 and 14,

References pp. 94–95

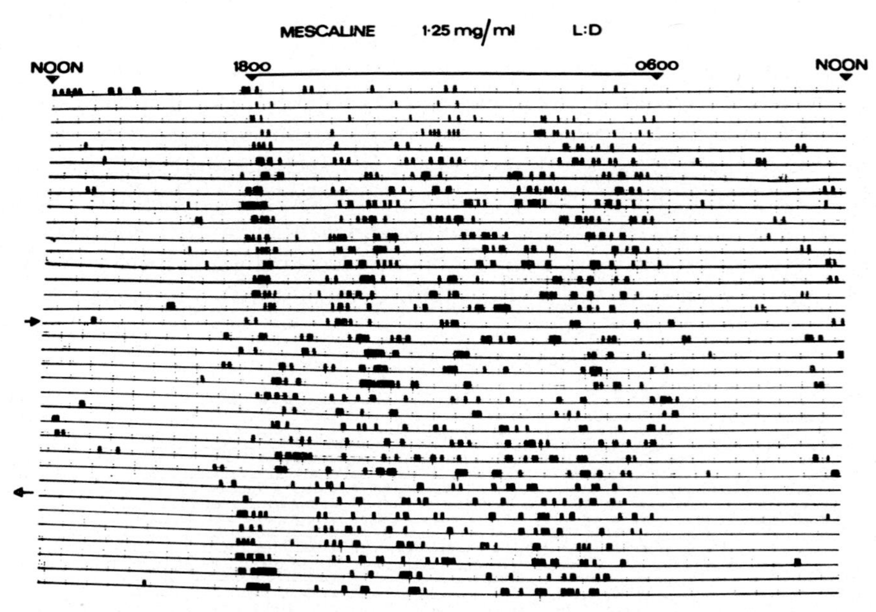

Fig. 10. The effect of mescaline (1.25 mg/ml) on the 24-h rhythm of locomotor activity in the rat.

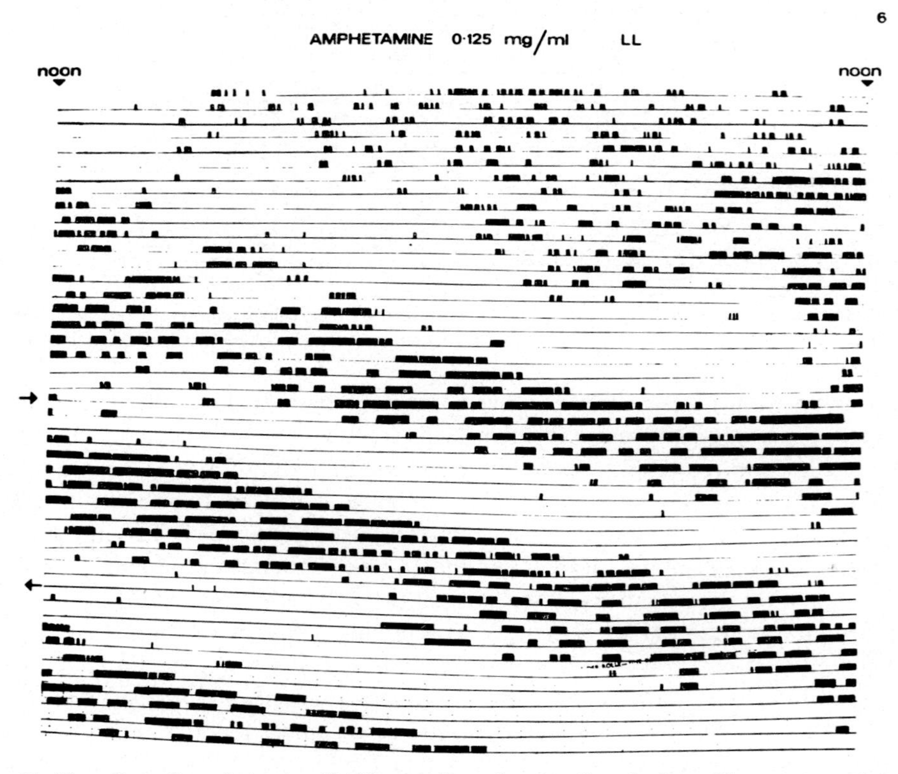

Fig. 11. The effect of amphetamine (0.125 mg/ml) on the circadian rhythm of locomotor activity in the rat, under continuous illumination.(→) Introduction of drug solution;(←) removal of drug solution.

in that although the amount of activity and the length of the activity period during the control period were quite different, the response to the drug was the same.

Continuous dark

Figs. 15 and 16 show the effect of amphetamine (0.25 mg/ml) on the circadian rhythms of rats in continuous darkness. It will be noticed that the deviation in the rhythm length from 24 h during the control period is considerably less than that found under continuous light (*cf.*, for instance, Fig. 11). Indeed, according to Aschoff (1963) the rhythm in continuous darkness should characteristically have a frequency of less than 24 h. This was not always so in our experiments, the frequency range being from 23 h 47 min to 24 h 10 min. It is also interesting to note (Fig. 15) the effect of including a female rat in an otherwise all-male experiment; the influence of the 4-day ovarian cycle on the circadian rhythm can be clearly seen.

Finally it should be noted that the effect of amphetamine on the circadian rhythm of rats kept in continuous darkness was considerably less than that on animals in continuous light. While it is generally held that circadian rhythms are more easily disturbed than synchronised rhythms, we know of no reason why the dark-oriented

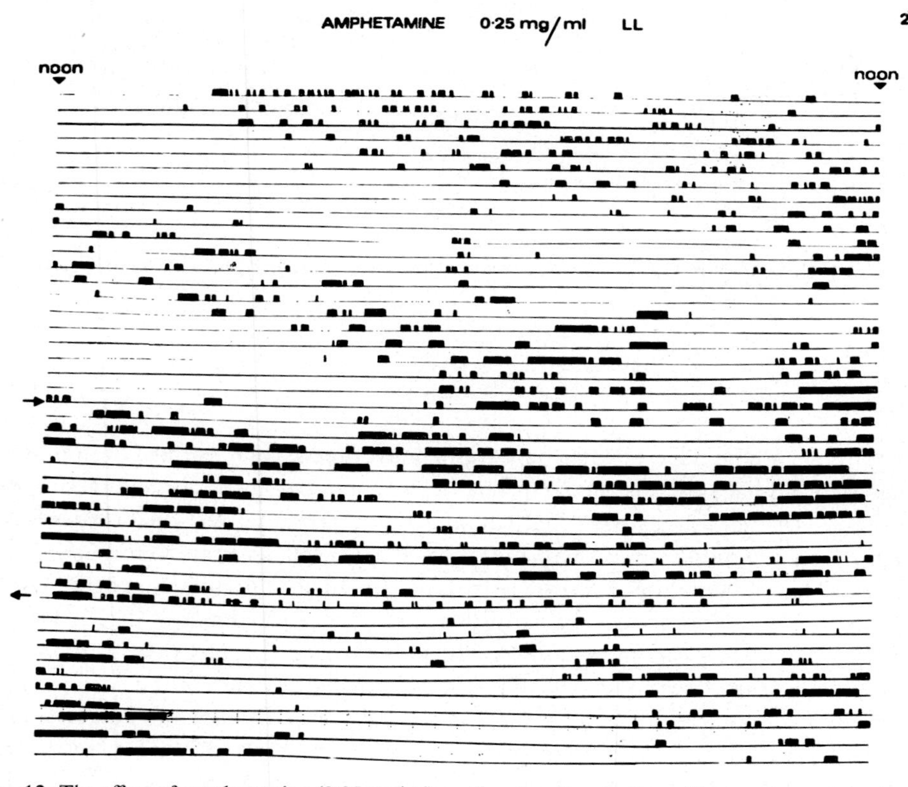

Fig. 12. The effect of amphetamine (0.25 mg/ml) on the circadian rhythm of locomotor activity in the rat under continuous illumination.(→) Introduction of drug solution;(←) removal of drug solution.

References pp. 94–95

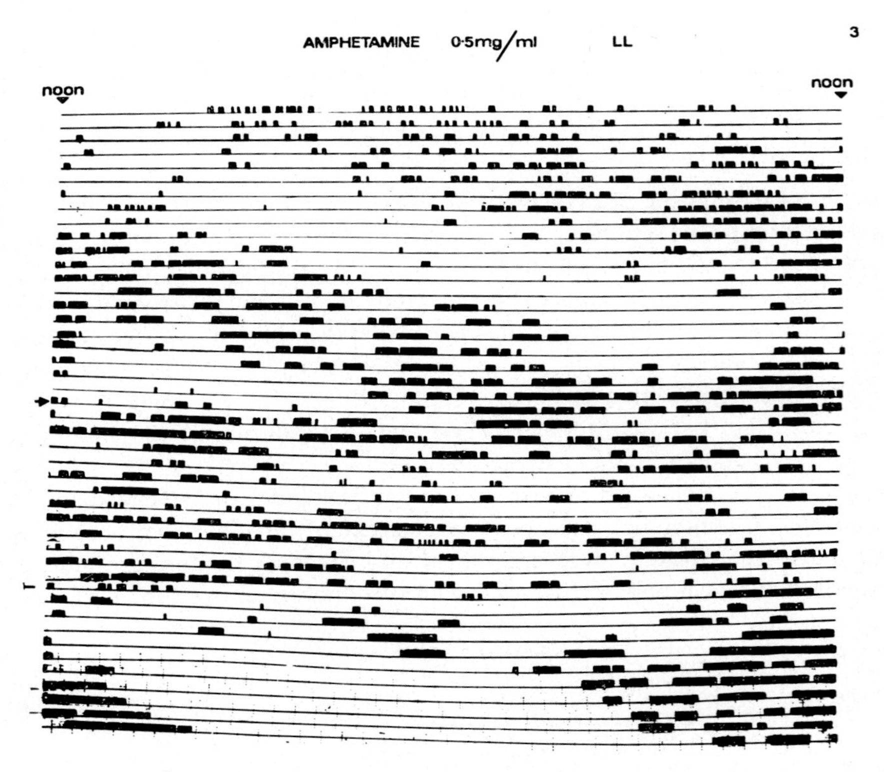

Fig. 13. The effect of amphetamine (0.5 mg/ml) on the circadian rhythm of locomotor activity in the rat under continuous illumination.(→) Introduction of drug solution; (←) removal of drug solution.

circadian rhythm should be more resistant to drug action than the light-orientated rhythm.

CONCLUSIONS

From these preliminary results, our original hypothesis that the psychotomimetic drugs were primarily acting on the mechanisms controlling circadian rhythms appears untenable. Of the 3 drugs used, only amphetamine produced any distortion of 24-h rhythms and it would seem probable that this effect is related to the central stimulant action of amphetamine rather than to any specific "psychotomimetic effect". However, before any firm conclusion is drawn, two facts must be considered: first, in carrying out these experiments, it was difficult to ensure comparability of dosage between the drugs. Our initial calculations were made in terms of the potency ratio following acute administration, and could not take into account the actual volume drunk. Consequently, the actual doses of mescaline and LSD were lower than originally intended. Fischer (1966) attributed the psychotomimetic "time-contracting" state induced by psylocibin, LSD and mescaline to a "central sympathetic

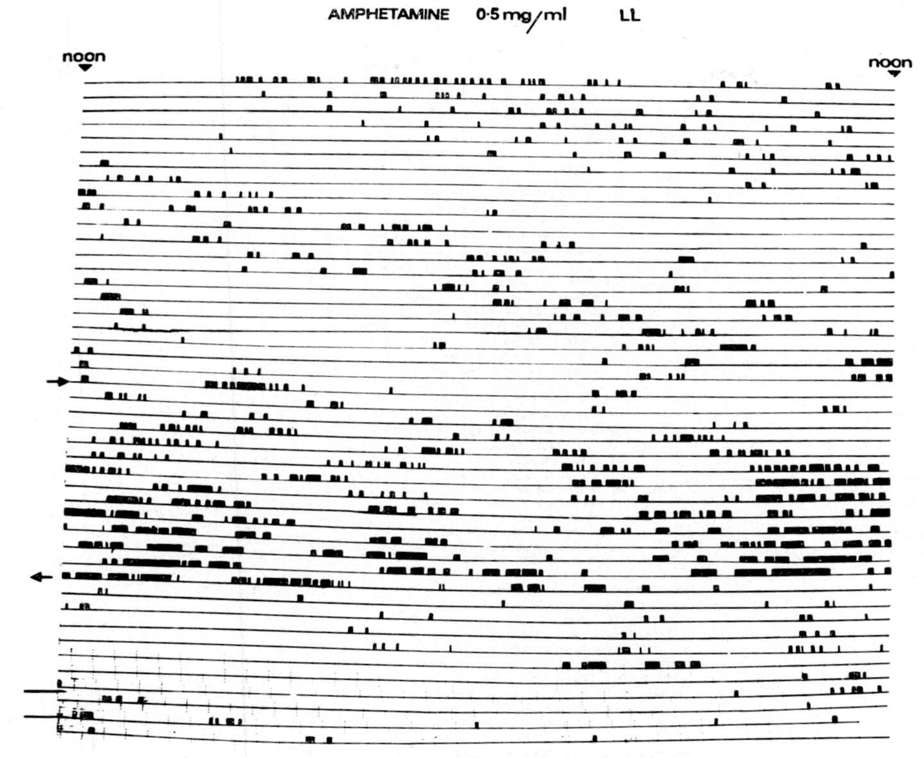

Fig. 14. See legend of Fig. 13.

excitement", and it may be that LSD and mescaline would have produced effects similar to those of amphetamine had higher doses been used.

Secondly, it is generally agreed that of all the "psychotomimetic" drugs, the toxic effects of amphetamine most closely resemble pathological psychosis in man. The possibility cannot be excluded, therefore, that the effects of amphetamine demonstrated here are related to the psychotomimetic action of the drug and may provide some clue to the cause of drug-induced and pathological psychosis in man.

References pp. 94–95

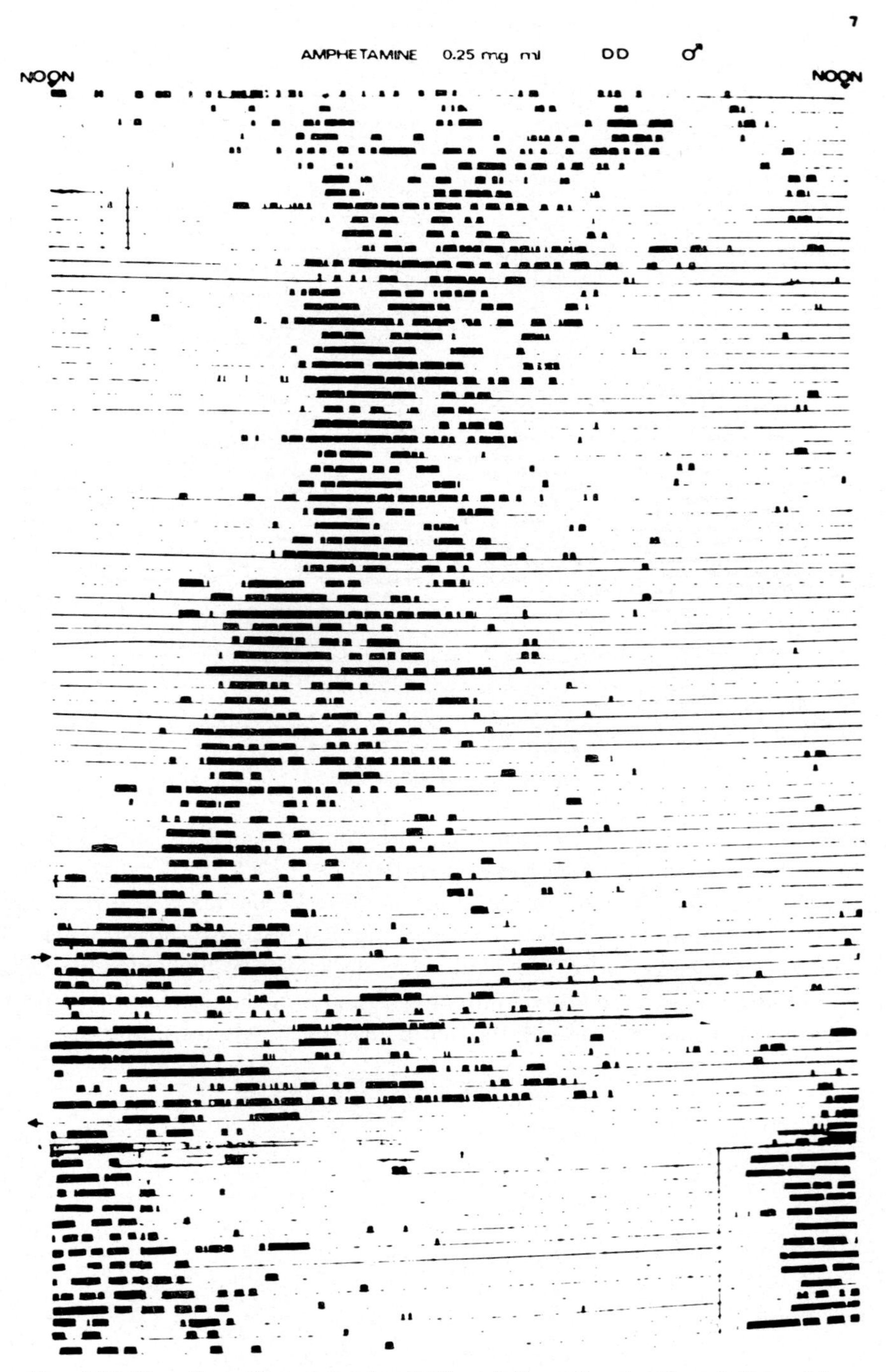

Figs. 15 and 16. The effect of amphetamine (0.25 mg/ml) on the circadian rhythm of locomotor activity in the rat in continuous darkness.(→) Introduction of drug solution;(←) removal of drug solution.

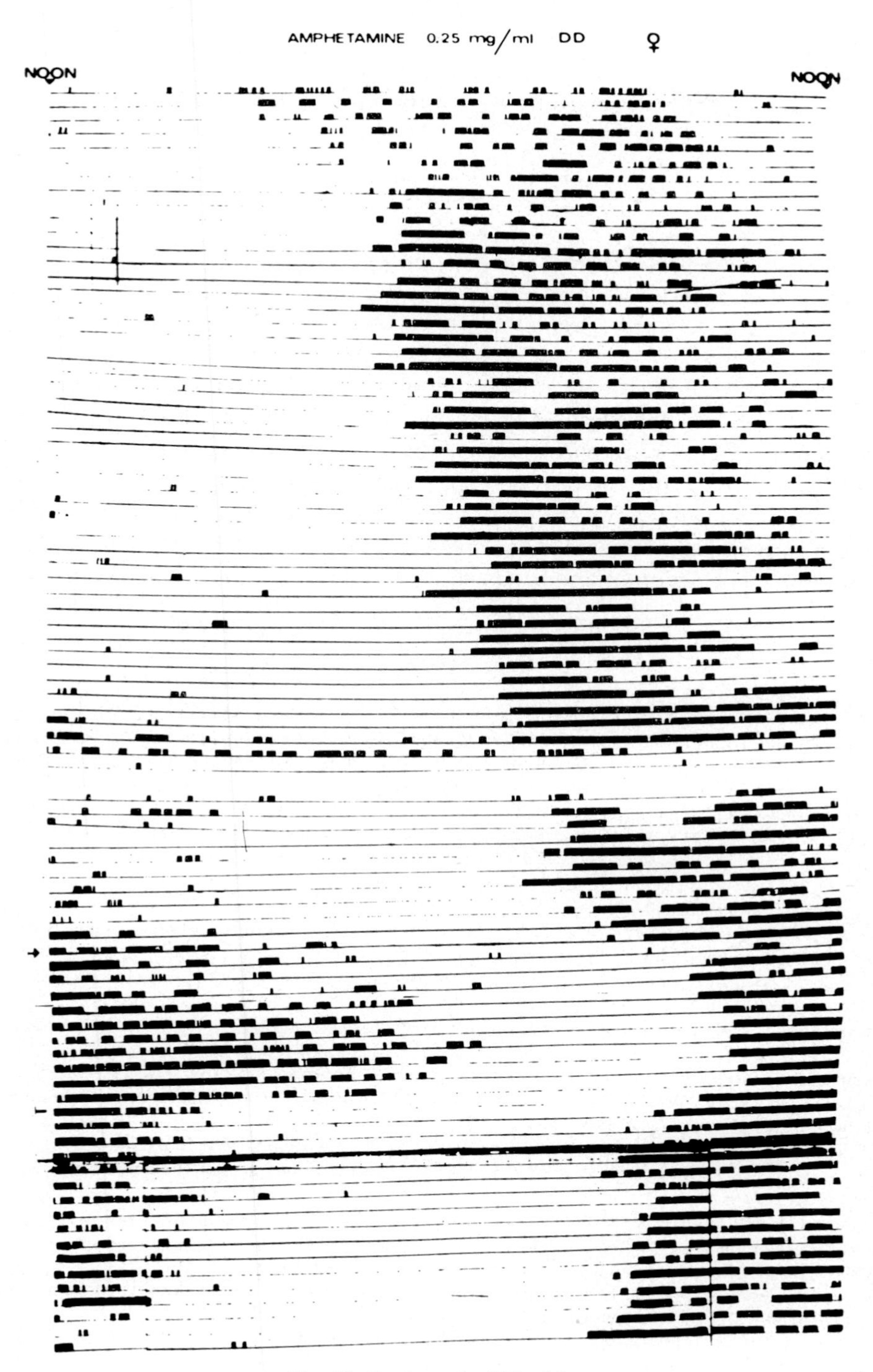

Fig. 16. See legend of Fig. 15.

References pp. 94–95

ACKNOWLEDGMENT

This work was partly supported by a grant from the Ministry of Defence.

SUMMARY

The interaction between psychotomimetic drugs and 24-h rhythms was investigated from two standpoints:

(1) The influence of clock-hour on the effect on locomotor activity of amphetamine, mescaline and LSD.

(2) The influence of chronic administration of these 3 drugs on 24-h rhythms.

It was shown that the effects of all 3 drugs were significantly affected by the clock-hour at which they were administered, and there was some indication that the variation might be related to 24-h rhythms in brain biogenic amine concentrations.

It was also shown that, of the 3 drugs, only amphetamine disrupted 24-h rhythms when chronically administered. The possibility that this effect might be related to the psychotomimetic action of amphetamine, and to psychosis in man, is discussed.

REFERENCES

Ancill, R. J., Davies, J. A. and Redfern, P. H. (1971) The influence of circadian rhythms on the effects of hallucinogenic drugs in the rat. In *Proc. of Syntex Symp.*, Suppl. to *J. clin. Trials*, in press.

Aschoff, J. (1963) Comparative physiology: Diurnal rhythms. *Ann. Rev. Physiol.*, **25**, 581–600.

Bebbington, A. and Brimblecombe, R. W. (1969) Actions of some toxic substances (psychotomimetics) on the central nervous system. *Brit. med. Bull.*, **26**, 293–298.

Cohen, M. and Wakely, H. (1968) A comparative behavioural study of ditran and LSD in mice, rats and dogs. *Arch. int. Pharmacodyn.*, **173**, 316–326.

Conroy, R. T. W. L., Elliot, A. L., Hughes, B. D. and Mills, J. N. (1968) Physiological variables in some psychotic patients. *J. Physiol. (Lond.)*, **196**, 129*P*–130*P*.

Del Rio, J. and Fuentes, J. A. (1969) Further studies on the antagonism of stereotyped behaviour induced by amphetamine. *Europ. J. Pharmacol.*, **8**, 73–78.

Dixit, B. M. and Buckley, J. P. (1967) Circadian changes in brain 5HT and plasma corticosterone in the rat. *Life Sci.*, **6**, 755–758.

Fischer, R. (1966) Sympathetic excitation and biological chronometry. *Int. J. Neuropsychiat.*, **2**, 116–121.

Fischer, R. (1967) The biological fabric of time. *Ann. N.Y. Acad. Sci.*, **138**, 440–488.

Friedman, A. H. and Walker, C. A. (1968) Circadian rhythms in rat mid-brain and caudate nucleus biogenic amine levels. *J. Physiol. (Lond.)*, **197**, 77–85.

Goldstein, A. (1964) *Biostatistics*: *An Introductory Text*, Macmillan, New York.

Lutsch, E. F. and Morris, R. W. (1967) Circadian periodicity to lidocaine hydrochloride. *Science*, **156**, 100–102.

Mathews, J. H., Marte, E. and Halberg, F. (1964) A circadian susceptibility resistance cycle to fluothane in male B1 mice. *Canad. Anaesth. Soc. J.*, **11**, 280–290.

Mills, J. M. (1966) Human circadian rhythms. *Physiol. Rev.*, **46**, 128–171.

Pauly, J. E. and Scheving, L. E. (1964) Temporal variations in the susceptibility of white rats to pentobarbital sodium and tremorines. *Int. J. Neuropharmacol.*, **3**, 651–658.

Quay, W. B. (1968) Individuation and lack of pineal effect in the rat's circadian locomotor rhythm. *Physiol. Behav.*, **3**, 109–118.

Richter, C. P. (1964) Biological clocks and the endocrine glands. In *Proc. 2nd. Int. Cong. Endocrinol., London*. Int. Congr. Series No. 83, Excerpta Medica, Amsterdam.

SCHEVING, L. E., HARRISON, W. H., GORDON, P. AND PAULY, S. E. (1968) Daily fluctuation in biogenic amines of the rat brain. *Amer. J. Physiol.*, **214**, 166–173.

DISCUSSION

KERKUT: Do you have information about the circadian levels of NA in animals kept under constant light or constant dark?

REDFERN: We have not studied this. As far as I am aware other people have examined only rhythms of NA in light–dark conditions.

KING: Have you considered whether activity changes after the chronic oral administration of amphetamine are attributable to the bitter taste of the drug?

REDFERN: This is always a possibility, as we did not attempt to mask the taste. However, when presented with the drug solution instead of water, the animals continued to drink the same volume per day.

BRADLEY: A simple explanation for the effects of amphetamine and the other two drugs in your experiments is that amphetamine is a central stimulant and not a psychotomimetic whereas the other two are.

REDFERN: Yes, I agree that is certainly one explanation. On the other hand amphetamine given in high doses does produce hallucinations and in that sense it has to be classed with the other two drugs.

ANSELL: You briefly mentioned the effect of seasonal variation on brain amines. How is the seasonal variation affected by laboratory conditions?

REDFERN: One would not expect any seasonal variation in controlled laboratory conditions. Nevertheless, a significant variation did occur in some of our experiments which were spread over 12 months.

BRIGGS: You showed that amphetamine was least effective on motor activity when the NA levels in the brains of the rats were highest: how do you interpret this in relation to the evidence that amphetamine acts by releasing NA?

REDFERN: Until we know why NA levels vary over 24 h we can only speculate. However, there are various ways in which one could explain the effect. For example, the increased level of NA may be a reflection of decreased availability for release.

BRIMBLECOMBE: It is well known that tolerance readily develops to the effects of LSD-25, and, I think, mescaline. I do not know whether this is the case with amphetamine, but presumably there is the possibility that had the LSD-25 been given in progressively increasing concentrations in the drinking water, some effect on the rhythm of spontaneous activity would have been seen.

REDFERN: Yes, this is possible. Unfortunately, with our existing experimental procedure, the animals have to be disturbed more often if increasing doses are to be given. We shall soon be able to introduce drug solution to the animals automatically from outside the environmental cabinet, and we should then be able to investigate this possibility.

WATTS: It is known that LSD-25 is unstable in aqueous solution unless great care is taken to exclude all traces of oxygen, so that the failure of your animals to exhibit any effects to LSD can probably be attributed to their not receiving any appreciable quantity of LSD.

REDFERN: It is certainly possible that some breakdown of LSD-25 may have occurred.

MEETER: Is there a way which makes it possible to decide whether the effects of amphetamine are really of central origin?

REDFERN: We have not, as yet, tried to seperate peripheral from central effects.

Effects of Some Centrally Acting Drugs on Caeruloplasmin

B. C. BARRASS AND D. B. COULT

Ministry of Defence, Chemical Defence Establishment, Porton Down, Salisbury, Wiltshire (Great Britain)

Potent centrally acting drugs are of importance to the biological scientist not only because of their effects on mood, perception and behaviour, but also because they can be used as specific chemical probes for studying the CNS at the neurophysiological and neurobiochemical levels. Before such drugs can be fully and rationally exploited for this purpose it is necessary that their mode of action should be understood in some detail; at the present time the mode of action of none of the centrally active drugs is sufficiently well understood for them to be so used. In view of their potent peripheral anticholinergic activity the glycollates (Abood, 1968), probably exert their central effects by interfering with central cholinergic pathways (Brimblecombe and Green, 1968), but the way in which compounds such as LSD, mescaline and 2,5-dimethoxy-4-methyl amphetamine (DOM, STP) produce their characteristic central effects is less well understood. In view of the chemical similarity (Table I) between LSD, mescaline and DOM on the one hand and central neurotransmitters such as dopamine, noradrenaline (NA) and 5-hydroxytryptamine (5-HT, serotonin) on the other, a mode of action involving interference with central dopaminergic, noradrenergic or serotoninergic mechanisms seems generally likely (Giarman and Freedman, 1965; Freedman and Aghajanian, 1966). Certainly, in the case of LSD its known peripheral antagonism of 5-HT (Gaddum, 1953) has been extensively quoted as supporting the suggestion that its central effects are due to some kind of interference with an unspecified central serotoninergic system. However, it must be admitted that attempts to account fully for the central effects of LSD, mescaline and similar compounds in terms of their interference with the synthesis, uptake, release, antagonism or destruction of natural neurotransmitters such as NA, dopamine and 5-HT have not so far proved entirely satisfactory.

In a search for a possible biochemical means of investigating the mode of action of LSD it was noted that the administration of this drug to animals resulted in an elevation of brain 5-HT levels and a depression of brain catecholamine levels (Barchas and Freedman, 1963; Koenig–Bersin *et al.*, 1970). It therefore seemed of interest to study the effects of LSD on an enzyme which could utilize both NA and 5-HT as substrates since it was considered that such an enzyme could be used as a model for those central receptors with which LSD must interact in order to produce its characteristic central effects. The enzyme which was eventually selected for study in this way was the copper-containing oxidase caeruloplasmin which utilizes both NA and

References p. 104

TABLE I

HALLUCINOGENIC DRUGS AND NEUROTRANSMITTERS

Hallucinogen	*Neurotransmitter*	*System Affected*
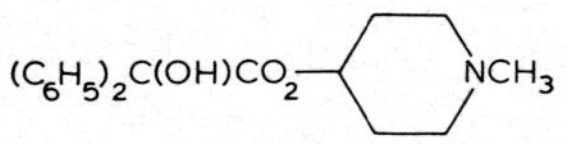 $(C_6H_5)_2C(OH)CO_2$–(piperidyl)–NCH_3	$CH_3CO_2CH_2CH_2N^+(CH_3)_3$	Cholinergic?
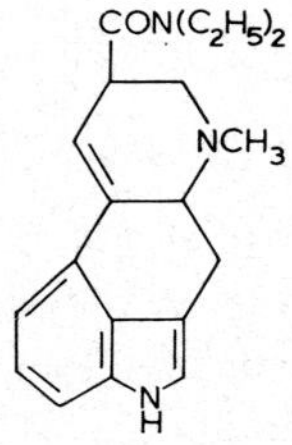 $CON(C_2H_5)_2$, NCH_3, N–H	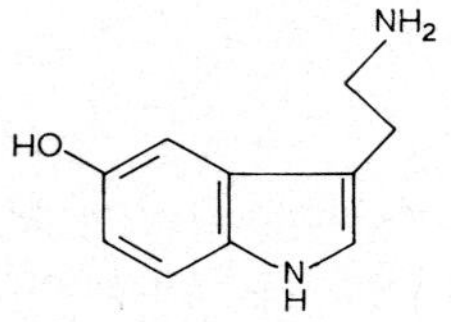 NH_2, HO, N–H	Serotoninergic?
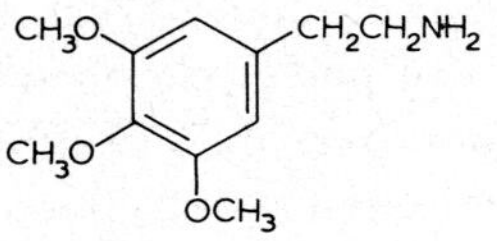 CH_3O, $CH_2CH_2NH_2$, CH_3O, OCH_3	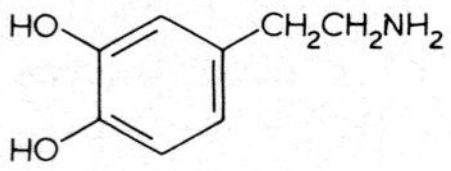 HO, $CH_2CH_2NH_2$, HO	Dopaminergic/ Adrenergic?
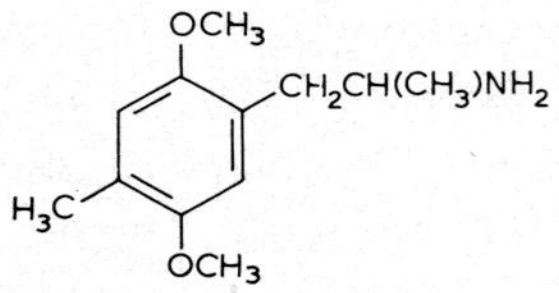 OCH_3, $CH_2CH(CH_3)NH_2$, H_3C, OCH_3	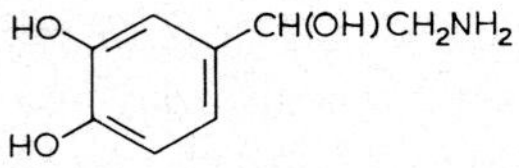 HO, $CH(OH)CH_2NH_2$, HO	

O_2

$4Cu^{2+}$

CP^{4Cu^+} + NA → CP^{8Cu^+} + FREE RADICAL

+ NADH

NAD^+

O, OH, $\ominus$O, $\oplus$N–H

NORADRENOCHROME

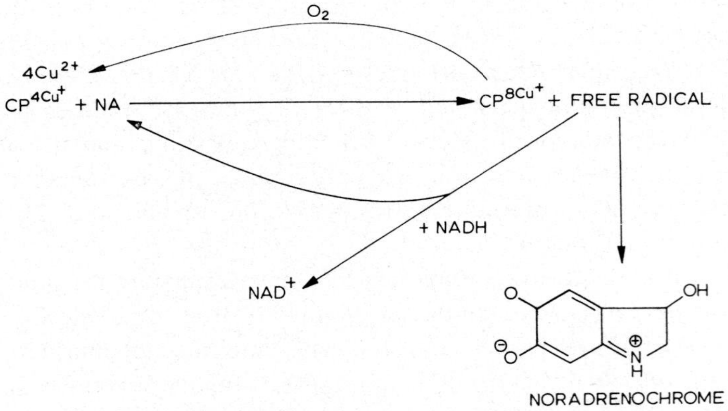

Fig. 1. *Mechanism of action of caeruloplasmin.*

$4Cu^{2+}$
CP^{4Cu^+}, Oxidised caeruloplasmin
CP^{8Cu^+}, Reduced caeruloplasmin
NA , Noradrenaline

5-HT as substrates in the cycle of reactions which are shown in Fig. 1. A similar scheme may be drawn using 5-HT as substrate.

The effects of various centrally active drugs on the caeruloplasmin-catalysed oxidation of NA and 5-HT were accordingly studied, the parameters monitored being the rates of formation of noradrenochrome and disappearance of NADH (both measured spectrophotometrically) and the rates of oxygen consumption and formation of NAD^+ (both being measured polarographically). The results obtained are summarised in Table II. Although all the compounds affected the caeruloplasmin-catalysed oxidation of both substrates, LSD had the most marked effect, in agreement with the fact that it is also the most potent hallucinogen of the 5 compounds shown in Table II.

TABLE II

EFFECTS OF SOME CENTRALLY ACTING INDOLES ON THE OXIDATION OF NORADRENALINE AND 5-HT BY CAERULOPLASMIN

Compound	*Action on oxidation of*		*Comments*
	Noradrenaline	*5-HT*	
LSD	Catalysis. 400% at a concentration 1/10 of substrate	Inhibition. 50% at a concentration 1/10 of substrate	Hallucinogen
Ibogaine	Catalysis. 200% at a concentration equal to substrate	Inhibition. 50% at a concentration equal to substrate	Hallucinogen
2-Bromo-LSD	Catalysis. 200% at a concentration ½ that of substrate	Inhibition. 50% at a concentration ½ that of substrate	Hallucinogen only at high doses
Harmine	Inhibition. 50% at 10^{-3}M	Inhibition. 50% at 10^{-3}M	Hallucinogenic activity not definitely established
Harmol	Inhibition. 50% at 10^{-4}M	Inhibition. 50% at 10^{-4}M	

The inhibition of the enzymic oxidation of 5-HT by LSD is not unexpected in view of the chemical similarities of the two molecules, but the acceleration of the enzymic oxidation of NA was totally unexpected and raises some interesting problems regarding the mechanism of action of caeruloplasmin. Perhaps the most likely explanation for these effects of LSD is that the enzyme has two distinct binding sites, for NA and 5-HT respectively, but that these have the oxidative site in common. LSD could then be considered to interact with the 5-HT binding site in such a way as to inhibit binding of this particular substrate to the enzyme and either enhance the binding of NA or enhance the chemical reactivity of the oxidative site. This suggestion that caeruloplasmin has two distinct active sites is supported by the work of Curzon and Speyer (1968) who, using inorganic inhibitors such as azide and

References p. 104

fluoride ions and unsymmetrical N,N-dimethyl *p*-phenylenediamine as the substrate, concluded that there were two sites at which these low molecular weight inhibitors could interact. An alternative possibility, that the effects of LSD are wholly allosteric in nature, cannot be ruled out at this time but there is no evidence that allosteric effects are involved. Further work is needed to clarify this point.

In view of the above results obtained with LSD it was clearly desirable to extend these studies to other centrally active drugs such as glycollates, phenylalkylamines and indoles. The glycollates described by Abood in 1968 were without effect on the caeruloplasmin-catalysed oxidation of NA and 5-HT, an observation which is in accord with the suggestion that they exert their central effects by interfering with central cholinergic pathways.

The phenylalkylamines mescaline, 2,5-dimethoxy-4-methyl amphetamine (DOM) and 3,4-dimethoxyphenylethylamine had no effects on caeruloplasmin — they were neither substrates nor did they modify the oxidation of known substrates. In other studies on the substrate specificity of caeruloplasmin it was noted that, in addition to being a substrate, 3-hydroxy-4-methoxyphenylethylamine accelerated the oxidation of both noradrenaline and 5-HT. Since demethylated metabolites of mescaline have been isolated by Musacchio and Goldstein (1967) from rats given this drug, it is possible that it is these metabolites which are the chemical species actually responsible for the observed central effects. In such a situation there should be a correlation between the central effects of the compound administered and the interaction of its demethylated metabolite with caeruloplasmin, but there would be no correlation between the central effects of the compounds and their effects on the enzyme. This point requires further experimental study.

The indoles described by Brimblecombe *et al.* (1964) had no effect on caeruloplasmin, but again it is possible that with these compounds it is the hydroxylated metabolites which are the active species, as suggested by the above authors and also by Szara (1961). These metabolites would, however, be formed by hydroxylation of the aromatic ring whereas the hydroxy metabolites of compounds such as mescaline and DOM would be formed by the demethylation of one of the methoxyl groups which are such a characteristic structural feature of the centrally active phenylalkylamines.

From the studies so far described it seemed that LSD was almost unique amongst centrally acting compounds in the potency of its effects on the caeruloplasmin-catalysed oxidation of NA and 5-HT, and this has led to the suggestion that this enzyme, or one with similar properties, may be directly involved in the mode of action of LSD. It is known that the K_m values for NA and 5-HT are almost identical, suggesting that caeruloplasmin or an enzyme with similar properties could exercise a very sensitive control over the relative concentrations of these two amines (and probably also dopamine) in those parts of the brain where they act as neurotransmitters. If, as seems likely, the maintenance of a balance between NA, dopamine and 5-HT is essential to normal mental function, then LSD could produce its central effects by disturbing the balance between these biogenic amines as a result of its interaction with caeruloplasmin or a similar enzyme; moreover, since LSD affects the

TABLE III

EFFECTS OF SOME PHENOTHIAZINES [phenothiazine ring structure with N–R and X substituent] ON THE OXIDATION OF NORADRENALINE AND 5-HT BY CAERULOPLASMIN

Compound		*Action on oxidation of*		*Comments*
R	*X*	*Noradrenaline*	*5-HT*	
$(CH_3)_2N(CH_2)_3$—	Cl	Catalysis. 200% at 10^{-3}M	Catalysis. 200% at 10^{-3}M	Tranquillizer (Chlorpromazine)
CH_3N[piperazine ring]$N(CH_2)_3$—	CF_3	Catalysis. 200% at 2 × 10^{-4}M	Catalysis. 200% at 2 × 10^{-4}M	Tranquillizer (Trifluperazine)
$(CH_3)_2N(CH_2)_3$—	CF_3	Catalysis. 200% at 10^{-3}M	Catalysis. 200% at 10^{-3}M	Tranquillizer (Triflupromazine)
[N-methylpiperidin-2-yl]$(CH_2)_3$—	SCH_3	Catalysis. 200% at 2 × 10^{-3}M	Catalysis. 200% at 2 × 10^{-3}M	Tranquillizer (Thioridazine)
$(CH_3)_2NCH(CH_3)CH_2$—	H	No effect	No effect	Anti-histaminic Anti-emetic (Promethazine)
$(C_2H_5)_2N(CH_2)_2$—	H	No effect	No effect	Anti-Parkinsonism (Diethazine)

References p. 104

TABLE IV

EFFECTS OF SOME ANTIDEPRESSANTS AND TRANQUILLIZERS ON THE CAERULOPLASMIN-CATALYSED OXIDATION OF NORADRENALINE AND 5-HT

Compound	*Effects on oxidation of*		*Comments*
	Noradrenaline	*5-HT*	
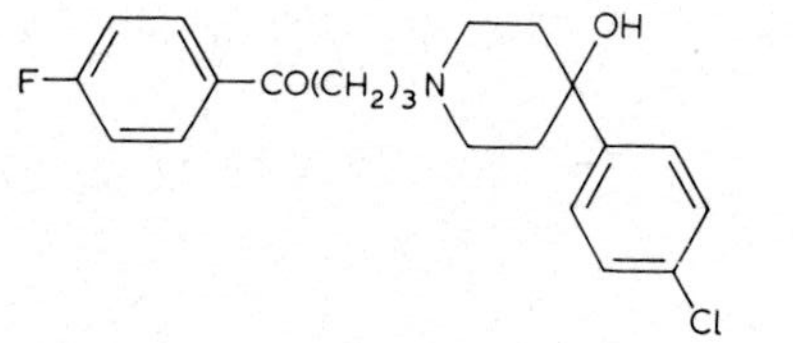 F, $CO(CH_2)_3N$, OH, Cl	Catalysis. 200% at 2×10^{-3}M	Catalysis. 200% at 2×10^{-3}M	Tranquillizer (Haloperidol)
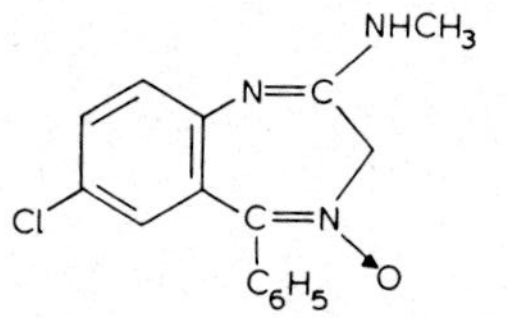 $NHCH_3$, N=C, C=N→O, Cl, C_6H_5	No effect	No effect	Minor tranquillizer (Chlordiazepoxide)
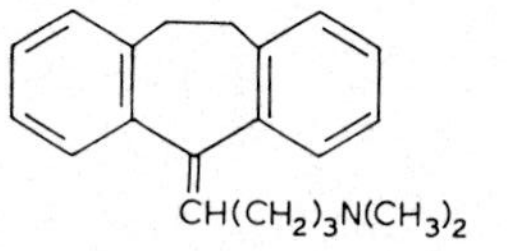 $CH(CH_2)_3N(CH_3)_2$	Inhibition. 50% at 10^{-2}M	Inhibition. 50% at 10^{-2}M	Anti-depressant (Amitryptyline)
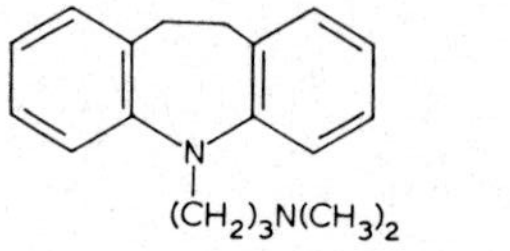 N, $(CH_2)_3N(CH_3)_2$	Inhibition. 50% at 10^{-2}M	Inhibition. 50% at 10^{-2}M	Anti-depressant (Imipramine)
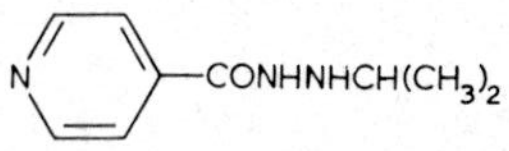 N, $CONHNHCH(CH_3)_2$	No effect	No effect	Anti-depressant (Iproniazid)

oxidation of NA and 5-HT in opposite directions it should be more potent than a compound which affects the oxidation of only one of these substrates. Implicit in this suggestion is the assumption that caeruloplasmin or a similar enzyme is directly involved in the maintenance of normal mental function. Since these investigations had been concerned only with compounds capable of modifying normal behaviour it was decided, as a result of the above considerations, to include in this study some compounds which are known to modify abnormal behaviour, *i.e.*, compounds used clinically for the treatment of abnormal mental states. Some of these results are shown in Tables III and IV. The tranquillizers of the phenothiazine type accelerated the oxidation of both NA and 5-HT, whereas phenothiazines which lack useful tranquillizing properties (promethazine and diethazine) had no effect. It is interesting that Haloperidol, which is not a phenothiazine but which is a tranquillizer, also accelerated the oxidation of both substrates, indicating that this effect is a reflection of tranquillizing properties and not of the phenothiazine structure *per se*. Chlordiazepoxide, a very weak tranquillizer used mainly in anxiety states rather than psychotic states, had no effect on caeruloplasmin.

The anti-depressants amitryptyline and imipramine, and their desmethyl analogues, inhibited the enzymic oxidation of both substrates but only at relatively high concentrations. It is significant that iproniazid, which owes its anti-depressant activity to its known inhibition of monoamine oxidase, had no effect on caeruloplasmin.

The above results tend to support the proposal that caeruloplasmin, or an enzyme with similar properties, plays an important role in the maintenance of normal mental function and that interference with this enzyme leads to the appearance of abnormal mental states. It is not possible at present, however, to draw any firm conclusions regarding the effects of a compound on caeruloplasmin and its potential clinical usefulness in the treatment of mental disorders. Such conclusions must await the results of further studies which the present work has shown to be desirable.

SUMMARY

1. The copper-containing oxidase caeruloplasmin was studied as a potential model for those receptors in the CNS with which certain centrally acting drugs must interact in order to produce their characteristic effects.

2. Lysergic acid diethylamide (LSD), and to a lesser degree ibogaine and 2-bromo-LSD, inhibited the enzymic oxidation of 5-HT but accelerated the oxidation of NA. Harmine and harmol inhibited the enzymic oxidation of both substrates.

3. Some phenylethylamines and anticholinergics with reported central activity had no effects on the enzymic oxidation of NA and 5-HT.

4. Certain of the drugs used in the treatment of mental illness affected the caeruloplasmin-catalysed oxidation of NA and 5-HT. Tranquillizers of the phenothiazine type, for example, accelerated the oxidation of both substrates, whilst anti-depressant drugs (other than monoamine oxidase inhibitors) inhibited the oxidation of both substrates.

5. The results have led to the tentative suggestion that caeruloplasmin, or an

References p. 104

enzyme with similar properties, may be of importance in controlling the relative concentrations of NA and 5-HT in some areas of the brain. The relevance of this suggestion to the mode of action of LSD and other centrally acting drugs is discussed.

REFERENCES

ABOOD, L. G. (1968) In *Drugs affecting the central nervous system.* Edward Arnold, London.

BARCHAS, J. D. AND FREEDMAN, D. X. (1963) Brain amines: response to physiological stress. *Biochem. Pharmacol.*, **12**, 1232–1235.

BRIMBLECOMBE, R. W., DOWNING, D. F., GREEN, D. M. AND HUNT, R. R. (1964) Some pharmacological effects of a series of tryptamine derivatives. *Brit. J. Pharmacol.*, **23**, 43–54.

BRIMBLECOMBE, R. W. AND GREEN, D. M. (1968) The peripheral and central actions of some anticholinergic substances. *Int. J. Neuropharmacol.*, **2**, 15–21.

CURZON, G. AND SPEYER, B. E. (1968) The effects of inhibitor mixtures and the specific effects of different anions on the oxidase activity of caeruloplasmin. *Biochem. J.*, **109**, 25–34.

FREEDMAN, D. X. AND AGHAJANIAN, G. K. (1966) Approaches to the pharmacology of LSD-25. *Lloydia*, **29**, 309–314.

GADDUM, J. H. (1953) Antagonism between lysergic acid diethylamide and 5-hydroxytryptamine. *J. Physiol. (Lond.)*, **121**, 15*P*.

GIARMAN, N. J. AND FREEDMAN, D. X. (1965) Biochemical aspects of the actions of psychotomimetic drugs. *Pharmacol. Rev.*, **17**, 1–25.

KOENIG–BERSIN, P., WASER, P. G., LANGEMANN, H. AND LICHTENSTEIGER, W. (1970) Monoamines in the brain under the influence of muscimol and ibotenic acid, two psychoactive principles of *Amanita muscaria. Psychopharmacologia (Berl.)*, **18**, 1–10.

MUSACCHIO, J. M. AND GOLDSTEIN, M. (1967) The metabolism of mescaline-^{14}C in rats. *Biochem. Pharmacol.*, **16**, 963–970.

SZARA, S. (1961) Correlation between metabolism and behavioural action of psychotropic tryptamine derivatives. *Biochem. Pharmacol.*, **8**, 32.

DISCUSSION

PINDER: Is dopamine a substrate for caeruloplasmin? If so, does the enzyme play a role in mediating dysfunctions of brain copper metabolism such as in Wilson's disease, particularly since dopaminergic deficiency is involved in the symptomatically related condition of Parkinsonism?

COULT: Yes, I have talked a great deal about noradrenaline, but when I said noradrenaline I could equally have said dopamine. The two seem very much the same as far as this enzyme is concerned. In fact, dopamine has a slightly higher affinity than noradrenaline.

KING: Is there any drug which gives inhibition of noradenaline and catalysis of 5-HT?

COULT: We do not know of any such compound.

BRIMBLECOMBE: I find your evidence concerning the psychotherapeutic drugs—the tranquillizers and anti-depressants—rather more convincing than the evidence relating to the psychotomimetic drugs. For example, the evidence for the 6-hydroxylation of tryptamine *in vivo* is not very strong and we would like to know that the hydroxylated mescaline derivative that you mentioned is a psychotomimetic drug.

COULT: These compounds must be active at the site. The fact that 6-hydroxytryptamines have been put into animals does not mean that these are the compounds which reach the active sites.

Some Biochemical Correlates of Inherited Behavourial Differences

J. T. RICK AND D. W. FULKER

Department of Psychology, University of Birmingham, Birmingham (Great Britain)

Behavioural and biochemical events are both continuous, ongoing activities. They are two functional aspects of any living system whose essential characteristic is one of change. In attempting to analyse how these two changing sets of events are interrelated, first the dynamics of the system must be placed under some form of adequate control. One way of achieving such control and coming to terms with this continuous stream of events is to exploit the stability of inherited differences in behaviour between strains of animals.

To appreciate how one can simultaneously control for the dynamic character of behaviour and biochemistry by means of the stability of inherited differences, let us first compare the ways of studying behaviour. The psychologist has two general approaches at his disposal for studying behaviour; either he analyses existing differences between two or more animals or he controls change in behaviour in individual animals. The use of genetics is a specific instance of the former approach, other such instances being environmentally contrived; these include brain lesions, imprinting and early handling: all of which result in animals which then continue to differ behaviourally from others which have not been so manipulated. Studies based on this differential approach measure behaviours which are essentially stable; that is for example, strains of animals, or early-handled and non-handled animals, differ from each other independently of temporal and environmental conditions. By contrast, studies based on behavioural change are essentially unstable, being dependent on temporal and environmental conditions. For example, behavioural change is commonly controlled either by drug administration or an imposed stimulus pattern. In the former case the animal's behaviour changes, peaks and then normally returns to the pre-drug state; in the latter the animal's behaviour having changed may well stabilize but this stability may wholly depend on the maintenance of environmental conditions as evidenced by the rapid behavioural change measured following the removal of the stimulus control during most extinction procedures (Rick, 1971). The instability of behavioural change becomes a severe constraint when attempting to relate biochemical and behavioural events; optimum behavioural change may not coincide with optimum biochemical change, as after the administration of reserpine (Robson and Stacey, 1962) or the imposed stimulus pattern controlling behavioural change may give rise to non-specific changes in pertinent biochemical mechanisms,

References pp. 111–112

as exemplified by the effects of electric shock on brain RNA synthesis which are independent of the pattern of the stimulation (Gardner *et al.*, 1970; Kerkut *et al.*, 1970). Clearly this constraint does not apply to studies based on behavioural differences. Further, when these differences are in terms of inherited characters we hope to show in this paper that genetic considerations allow one to relate more directly biochemical with behavioural activity.

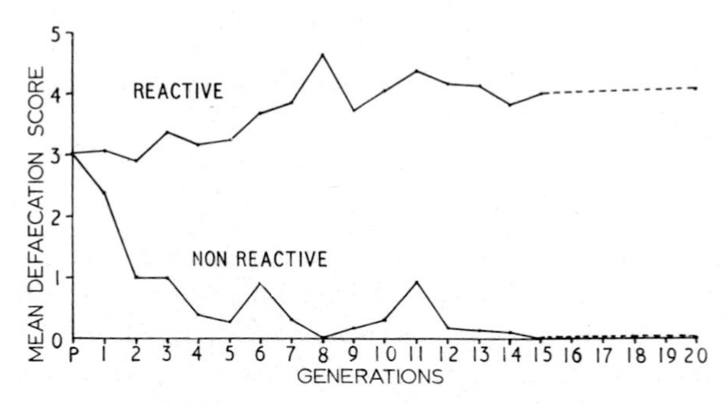

Fig. 1. Progressive selection for high and low defaecation by the rat in the open field. The ordinate shows the mean number of faecal boluses deposited per trial and the abscissa the scores of successive generations. Selection was suspended from the fifteenth to the twentieth generation when the measure showed no tendency to revert to the foundation population score.

Behavioural and biochemical differences may arise incidentally between specific strains (Al–Ani *et al.*, 1970) or may be deliberately bred for as can be seen in Fig. 1, which is taken from a selection experiment by Broadhurst (1967). Rats in this experiment were selected for high or low reactivity to mild stress, reactivity being measured by defaecation in the open field. It can be seen from Fig. 1 that selection pressure over a few generations establishes two behaviourally distinct strains of rat from a common stock, the Maudsley Reactive (MR) and the Maudsley Non-Reactive (MNR) strains. When selection was suspended from the fifteenth generation until the twentieth there was little evidence of reversion to the foundation population score and it has been argued elsewhere (Rick *et al.*, 1967) that such contrasting patterns of behaviour, which have been as it were genetically "frozen", reflect functional differences in brain metabolism. Rick *et al.* (1967) have shown that such rats bred for high and low defaecation under mild stress differ significantly in the production of γ-aminobutyric acid (GABA) in the sensori-motor cortex and, using the same behavioural parameter, Broadhurst and Watson (1969) showed a similar relationship between reaction to stress and cholinesterase (ChE) activity. Other investigations have indicated that brain ChE activity is positively correlated with glutamic acid decarboxylase (EC 4.1.1.15) activity, which is the enzyme concerned in the production of GABA and both of which systems are related to learning (Geller *et al.*, 1965; Oliver *et al.*, 1971). The possible relationship between ChE activity and GABA production has also been studied in an experiment based on the stability of inherited differences (Rick *et al.*, 1968). Five strains of rat were used, 4 of which had been bred specifically for learning ability. These comprised the Tryon Maze Dull (TMD), Tryon Maze Bright (TMB), Roman High Avoidance (RHA) and the Roman

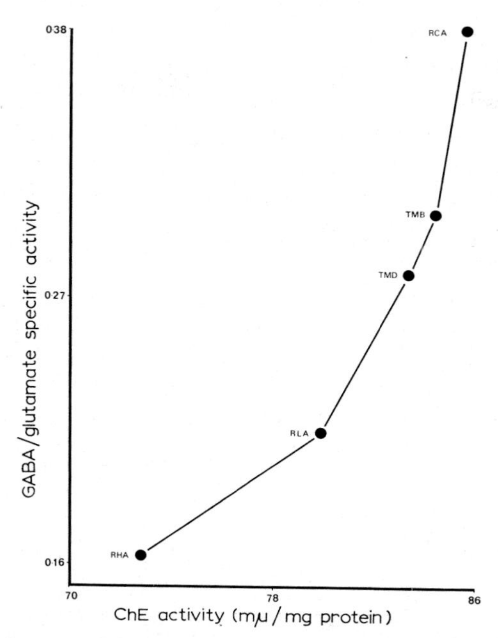

Fig. 2. The relationship between GABA production and ChE activity in the sensori-motor cortex of 5 strains of rat. (See Table II for full names of all rat strains.)

Low Avoidance (RLA). The fifth strain was an unselected strain of the original Roman stock, the Roman Control Avoidance (RCA) strain. As can be seen from Fig. 2, by making use of discrete behavioural strains and thus increasing the experimental variance while holding the error variance constant, a clear correlation is apparent between the two characters. This study has since been replicated on 8 strains in which the correlation between GABA production and ChE activity in the sensori-motor cortex was +0.88 ($P < 0.001$).

Obtaining such interrelated differences between strains of animals is obviously only the first step. One has then systematically to alter one character and look for correlated change in the other. This can be done in two ways, either in terms of behavioural change or by using more sophosticated genetic experiments. Both approaches are of value and we shall give an example of the behavioural change technique first from a series of ongoing experiments with *Drosophila melanogaster*.

Connolly (1966) has selectively bred *Drosophila* for level of spontaneous locomotor activity. The directional selection imposed resulted in an extremely inactive line and a highly active one. It has been established by Connolly (1967) that the principal phenotypic difference between the strains was spontaneous activity and not reactivity to environmental stimuli. The genetic system involved in the expression of phenotype is polygenic and the character has a heritability of 0.51 ± 0.10. The concentrations of biogenic amines were assayed in the two selected strains and in an unselected, control strain (Tunnicliff *et al.*, 1969). Dopamine concentrations were maximal in the inactive line and lowest in the active, the control strain having an intermediate dopamine

References pp. 111–112

level. Having obtained this systematic difference in dopamine levels between the 3 lines, we then manipulated the levels of dopamine in the active and control strains by rearing the flies on a medium containing gamma-hydroxybutyric acid (GHB) (Connolly *et al.*, 1971). Following the administration of GHB only dopamine, of the biogenic amines, is significantly affected in nervous tissue (Gessa *et al.*, 1968), when it increases up to 3 times its normal concentration. At the dose levels used, shown in Table I, GHB increased dopamine levels nearly 100% in the active strain, but did not

TABLE I

THE EFFECTS OF γ-HYDROXYBUTYRIC ACID IN CONCENTRATION OF 0.75% (BY WEIGHT) ON SPONTANEOUS ACTIVITY AND DOPAMINE LEVELS IN A SELECTED AND CONTROL LINE OF *Drosophila melanogaster*. THE VALUES ARE MEANS ± S.E.M. OF 20 ESTIMATIONS OF ACTIVITY AND 3 ESTIMATIONS OF DOPAMINE LEVEL.

		Medium alone	*Medium plus GHB*
Active line	Activity	61.5 ± 3.5	14.0 ± 2.3
	Dopamine	0.98 ± 0.25 μg/g	1.56 ± 0.06 μg/g
Control line	Activity	38.2 ± 7.6	39.1 ± 8.0
	Dopamine	1.09 ± 0.14 μg/g	0.84 ± 0.48 μg/g

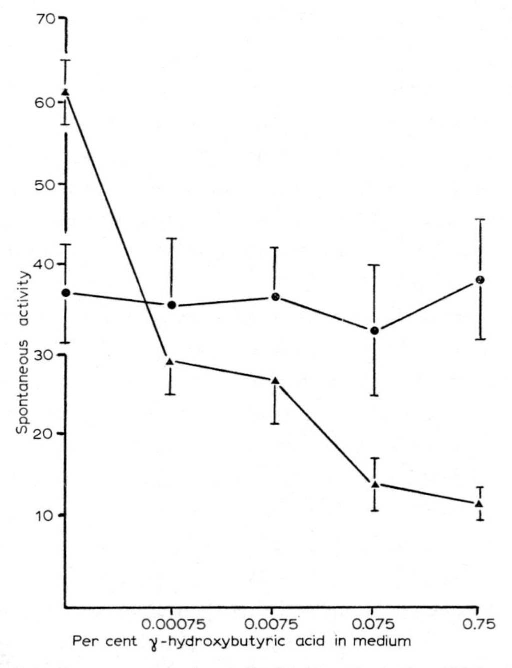

Fig. 3. Effects on increasing the concentration of γ-hydroxybutyric acid in the nutrient medium on the spontaneous locomotor activity of *Drosophila melanogaster*. The active strain (▲—▲) exhibited spontaneous activity which decreased in inverse proportion to the amount of drug in the medium. The unselected control strain (●—●) was not so affected over the dose range used. Each point is the mean activity of 20 flies at each drug condition, S.E.M.'s are indicated by bars.

systematically affect the concentration of dopamine in the control strain. The behavioural effects of the drug were consistent with these biochemical changes and the dose–response curves for both strains are shown in Fig. 3. The lack of effect in the control strain at the dose levels used is taken to reflect its greater heterozygosity and therefore buffering capacity against the perturbations of the environment, compared to the selected strain which had reached the one hundred and twenty-eighth generation of directional selection at the time of the experiment.

The genetic system involved in nearly all the behaviours upon which differential studies between strains are based is polygenic. That is, the expression of the behavioural phenotype is controlled not by single genes but by many. By exploiting this polygenic determination, a direct interrelationship between behavioural and biochemical characters may be defined. In this approach the change in phenotypic expression by environmental manipulation, described earlier using *Drosophila*, is replaced by genetic manipulation made possible by the segregation of the genes following imposed cross-breeding. Since n genes can give rise to 3^n different genotypes, with as few as 3 genes underlying both the biochemical and behavioural systems as many as 27 different levels of activity can be established by the appropriate techniques of breeding. In effect we have replaced the comparison of the genotypes represented by a number of strains with the comparison of a much greater number of genotypes created by experimental breeding. A number of advantages over the environmental manipulation of strains follow. Firstly, the greater number of comparisons involved allow any relationship to be more clearly established. Secondly, the variety of specified genotypes allows more complex relations to be investigated. Thirdly, these genotypes are more like those arising in a free-mating natural population. Fourthly, provided that appropriate breeding designs are used, segregation breaks up fortuitous relationships between measures that may have been present in the original strains.

Little has been done to exploit polygenic pleiotropy as a means of interrelating biochemistry and behaviour. One study where the pleiotropic approach has been used to investigate neurochemical correlates of behaviour was based on the differences in GABA production in the cortex of 5 strains of rats referred to earlier (Fig. 2). Eight strains of rats formed the basis of the study (Rick *et al.*, 1971). These were TMD, TMB, RHA, RLA and RCA, as previously mentioned, and the MNR and MR strains (Eysenck and Broadhurst, 1964) as well as the Wistar Albino Glaxo (WAG) strains. A diallel cross design in which each strain is crossed with every other, based on the approach of Broadhurst and Jinks (1966), was used. Behavioural measures derived from the open-field test and shock avoidance conditioning were made, as well as GABA production in the sensori-motor cortex. Half the rats in the study were handled daily during the pre-weaning period of 21 days. A variety of genetic influences and interactions with the handling were found (Fulker *et al.*, 1972; Fulker and Rick, 1972). In particular, activity measures and GABA production followed broadly similar patterns in which F_1 rats were buffered by dominant genes against the disruptive effects of handling. Moreover the two kinds of measurement, biochemical and behavioural, were inversely correlated in the strains. A preliminary analysis of this pleiotropy was carried out using additive genetic correlations based on the array

TABLE II

ARRAY MEANS OF AN 8 × 8 DIALLEL CROSS FOR 5 MEASURES

Strain	*Measures*				
	GABA Production	*Avoidances*	*ITC*	*Ambulation*	*Defaecation*
Roman High Avoidance (RHA)	8.54	13.55	4.65	8.48	1.00
Roman Low Avoidance (RLA)	9.59	7.30	3.39	6.85	0.90
Roman Control Avoidance (RCA)	9.77	10.28	3.32	8.01	0.62
Maudsley Reactive (MR)	9.28	12.03	3.92	7.48	0.68
Maudsley Non-Reactive (MNR)	8.73	13.49	4.55	8.42	0.22
Tryon Maze Bright (TMB)	8.99	10.79	4.09	6.46	0.84
Tryon Maze Dull (TMD)	9.33	9.36	3.61	7.07	0.52
Wistar Albino Glaxo (WAG)	8.47	14.63	5.35	8.62	0.40
Standard error*	±0.24	± 0.34	±0.08	±0.08	±0.02

The units are: GABA production, ^{14}C incorporation ratio for GABA/glutamate from U-[^{14}C] glucose, *in vitro* × 100; avoidances, number of avoidances out of 30 trials in escape–avoidance conditioning; ITC, intertrial crossings during escape–avoidance conditioning (transformed to square root); ambulation, number of metres run on 4 days of testing in the open field; and defaecation, number of faecal boluses deposited in open field on 4 days of testing (transformed to square root).

* A pooled S.E.M. for each measure is appropriate in this design.

TABLE III

CORRELATION MATRIX OF ARRAY MEANS IN TABLE II

	2	*3*	*4*	*5*
1 GABA production	—0.85	—0.96	—0.55	0.20
2 Avoidances		0.90	0.79	—0.38
3 ITC			0.65	—0.32
4 Ambulation				—0.41
5 Defaecation				—

means of the diallel table (Rick *et al.*, 1971). The means are given in Table II, and Table III shows the correlations between these 5 measures. Inspection of Table III indicates that GABA production, avoidance and intertrial crossing are all highly intercorrelated in one biological system and that GABA production is weakly correlated to emotional reactivity as measured by defaecation in the open-field test. A principal component analysis of the correlation matrix confirms this. Two orthogonal components were identified which account for 89% of the total observed variance. The first component involved a learning–activity dimension and identified with GABA production ($r = -0.95$, $P < 0.001$) but had a zero loading for the defaecation measure. The second identified with defaecation ($r = +0.97$, $P < 0.001$) and had a negligible loading for GABA production ($r = 0.14$). The overall results of this analysis are given in Table IV. Thus by selective cross-breeding a pleiotropic relationship

TABLE IV

FIRST TWO PRINCIPAL DIMENSIONS FOR CORRELATION MATRIX IN TABLE III

Component	*GABA production*	*Avoidances*	*ITC*	*Ambulation*	*Defaecation*	*Per cent of variance accounted for*
I	—0.95	0.87	0.93	0.62	0	58
II	0.14	—0.42	—0.28	—0.57	0.97	31

I and II are the first principal dimensions of the matrix in Table III. The dimensions are orthogonal and have been rotated through 20 ° to give the zero loading shown for defaecation on dimension I.

between GABA production in the sensori-motor cortex and a component of the rat's behaviour is strongly suggested.

In conclusion, we have tried to show that one can control for the dynamic interrelationship between behaviour and biochemistry by exploiting the stability of inherited differences between strains of animals. Further by analysing such genetic material quantitatively it is possible to establish pleiotropic relationships between biochemical and behavioural characters.

SUMMARY

If attempts to relate biochemical with behavioural events are to be meaningful, the dynamic character of the two sets of events must be adequately controlled for. A powerful way of obtaining such control is to exploit the stability of inherited differences between strains of animals. Data from a number of experiments illustrating this genetic approach are described. Further it is argued that biochemical events can be directly related to behavioural activity with the use of biometrical genetics.

ACKNOWLEDGEMENTS

We gratefully acknowledge financial support from the Mental Health Research Fund (J.T.R.) and the Medical Research Council (D.W.F.).

REFERENCES

AL-ANI, A. T., TUNNICLIFF, G., RICK, J. T. AND KERKUT, G. A. (1970) GABA production, acetylcholinesterase activity and biogenic amine levels in brain for mouse strains differing in spontaneous activity and reactivity. *Life Sci.*, **9**, 21–27.

BROADHURST, P. L. (1967) The biometrical analysis of behavioural inheritance. *Sci. Prog.*, **55**, 123–139.

BROADHURST, P. L. AND JINKS, J. L. (1966) Stability and change in the inheritance of behaviour in rats: a further analysis of statistics from a diallel cross. *Proc. roy Soc. B*, **165**, 450–472.

BROADHURST, P. L. AND WATSON, R. H. J. (1969) Brain cholinesterase, body build and emotionality in different strains of rats. *Anim. Behav.*, **12**, 42–51.
CONNOLLY, K. (1966) Locomotor activity in *Drosophila*. II. Selection for active and inactive strains. *Anim. Behav.*, **14**, 444–449.
CONNOLLY, K. (1967) Locomotor activity in *Drosophila*. III. A distinction between activity and reactivity. *Anim. Behav.*, **15**, 149–152.
CONNOLLY, K., TUNNICLIFF, G. AND RICK, J. T. (1971) The effects of gamma-hydroxybutyric acid on spontaneous locomotor activity and dopamine level in a selected strain of *Drosophila melanogaster*. *Comp. Biochem. Physiol.*, **408**, 321–326.
EYSENCK, H. J. AND BROADHURST, P. L. (1964) Experiments with animals: Introduction. In *Experiments in Motivation*, H. J. EYSENCK (Ed.), Pergamon Press, Oxford, pp. 285–291.
FULKER, D. W. AND RICK, J. T. (1972) The inheritance of GABA production in rat cortex. Unpublished data.
FULKER, D. W., WILCOCK, J. AND BROADHURST, P. L. (1972) Studies in genotype environment interaction. I. Methodology and preliminary multivariate analysis of variance of a diallel cross of eight strains of rat. *Behav. Genet.*, in press.
GARDNER, F. T., DEBOLD, R. C., FIRSTEIN, W. AND HEERMANS, JR., H. W. (1970) Increased incorporation of ^{14}C-uridine into rat brain RNA as a result of novel electric shock. *Nature (Lond.)*, **227**, 1242–1243.
GELLER, E., YUWILER, A. AND ZOLMAN, J. F. (1965) Effects of environmental complexity on constituents of brain and liver. *J. Neurochem.*, **12**, 949–955.
GESSA, G. L., CRABAI, F., VARGIU, L. AND SPANO, P. F. (1968) Selective increase of brain dopamine induced by gamma-hydroxybutyrate: study of the mechanism of action. *J. Neurochem.*, **15**, 377–381.
KERKUT, G. A., OLIVER, G. W. O., RICK, J. T. AND WALKER, R. J. (1970) The effects of drugs on learning in a simple preparation. *Comp. gen. Pharmacol.*, **1**, 437–483.
OLIVER, G. W. O., TABERNER, P., RICK, J. T. AND KERKUT, G. A. (1971) Changes in GABA level, GAD and ChE activity in the CNS of an insect during learning. *Comp. Biochem. Physiol.*, **38**, 529–535.
RICK, J. T. (1971) Neurochemical correlates of behavioural change: a problem in dynamics. In *Macromolecules and Behaviour*, E. B. Ansell and P. B. Bradley (Eds.), Macmillan, London. In press.
RICK, J. T., FULKER, D. W., TUNNICLIFF, G., WILCOCK, J., KERKUT, G. A. AND BROADHURST, P. L. (1971) GABA production in brain cortex related to activity and avoidance behaviour in eight strains of rat. *Brain Res.*, **32**, 234–238.
RICK, J. T., HUGGINS, A. K. AND KERKUT, G. A. (1967) The comparative production of gamma-aminobutryic acid in the Maudsley Reactive and Non-Reactive strains of rat. *Comp. Biochem. Physiol.*, **20**, 1009–1013.
RICK, J. T., MORRIS, D. AND KERKUT, G. A. (1968) Cholinesterase, cholineacetyltranferase and the production of gamma-aminobutyric acid in the cerebral cortex of five behavioural strains of rat. *Life Sci.*, **7**, 733–739.
ROBSON, J. M. AND STACEY, R. S. (1962) *Recent Advances in Pharmacology*. Churchill, London.
TUNNICLIFF, G., RICK, J. T. AND CONNOLLY, K. (1969) Locomotor activity in *Drosophila*. V. A comparative study of selectively bred populations. *Comp. Biochem. Physiol.*, **29**, 1239–1245.

DISCUSSION

FONNUM: Can you tell us which step in GABA synthesis is responsible for the changes you observe? Is it the uptake of glutamate into nerve terminals, activity of glutamate decarboxylase, or what?

RICK: We did not change the production of GABA in the animals, but measured the differences that already existed between the strains. The ratio of radio carbon from U-[^{14}C]glucose into GABA compared to that in glutamate is a measure of GABA synthesis and therefore, indirectly, of the optimal, rather than maximal, rate of glutamate decarboxylase activity for the strains.

CREASEY: In selecting your animals for different behavioural characteristics you take no account of changes in brain metabolites other than GABA. Is it therefore justifiable to attribute the behavioural changes that you have observed solely to changes in GABA metabolism? Is it not possible that you may have missed the metabolite of real significance through not looking for it.

RICK: Yes, this is possible or even likely. The point is that in this sort of work you have to find substances in brain which may be of importance and that you can get at reliably and consistently. We assayed for AChE, dopamine, 5-HT and NA as well as GABA. We used this particular way of assaying for GABA production because it was simple, straightforward and consistent. The data as presented indicate that GABA is highly identified with this arousal parameter. That biochemical mechanisms in the brain are likely to be equally, if not more so identified I would not dispute.

KING: You spoke about the level of GABA being related to the amount of activity of the animals; is it not possible that some of these animals may just be sick and therefore do not move around as much as other animals and that this accounts for the behavioural differences you have observed, and also for the differences in their brain GABA levels.

RICK: No. It is conceivable that one or two strains, by chance or by some artifact, may be sick and that their behaviour is a concomitant of this rather than of some aspect of their brain chemistry. If this were so, however, one would pick it up because all the strains would have to be sick in the same way in order to get this type of identification on the component. I am not terribly happy with the term "arousal". I tend to think of it more as the state of reactivity of the animal, but this tends to get confused with emotional reactivity.

KING: I thought from your previous work that you were not talking about reactivity but about levels of spontaneous activity.

RICK: No. Up to now we have looked at spontaneous activity only in *Drosophila*. Connolly (1967) distinguished between this phenomenon and reactivity to environmental stimulation.

CREASEY: Do you agree that half the total metabolism of pyruvate goes through the GABA shunt, and if it does, do you think that it is reasonable to suggest as an important transmitter a compound having such a vigorous metabolic role?

RICK: Recent data have shown that it is about 8%, if that, going through the GABA shunt (Balázs *et al.*, 1970). Even this is high and it seems to occur all through the brain whereas one of the things we like about a transmitter substance is that it is high in some parts of the brain and low in others. But that is only because we have some transmitter substances which are distributed like that, and we cannot demote something because it does not fit into that particular pattern. Secondly, if the heterogeneity of glutamate metabolism is considered, there would seem to be a large pool more associated with glutamine metabolism and a small pool more associated with GABA metabolism, but the actual variation between these two pools is not entirely specific and I would argue that in all probability you can demonstrate GABA production throughout the brain but that in some parts it is not acting as a transmitter—that is, it is not active in those pools which are associated with transmission.

SZERB: Your observation on increased GABA production in the less aroused strains of rats are supported by Jasper's findings on increased release of GABA during low-frequency, high-voltage EEG activity (Jasper *et al.*, 1965).

RICK: There is also the work of Mitchell and Srinivasan (1969) in which the inhibitory input to the cortex is stimulated to give an increase in the amount of GABA released.

REFERENCES

BALÁZS, R., MACHIYAMA, Y., HAMMOND, B. J., JULIAN, T. AND RICHTER, D. (1970) The operation of the γ-aminobutyrate bypath of the tricarboxylic acid cycle in brain tissue *in vitro*. *Biochem. J.*, **116**, 445–460.

CONNOLLY, K. (1967) Locomotor activity in *Drosophila*. III. A distinction between activity and reactivity. *Anim. Behav.*, **15**, 149–152.

JASPER, H. H., KAHN, R. T. AND ELLIOTT, K. A. C. (1965) Amino acids released from the cerebral cortex in relation to its state of activation. *Science*, **147**, 1448–1449.

MITCHELL, J. F. AND SRINIVASAN, V. (1969) Release of ^{3}H-γ-aminobutyric acid from the brain during synaptic inhibition. *Nature (Lond.)*, **224**, 663–666.

Behavioural Actions of Anticholinergic Drugs

R. W. BRIMBLECOMBE AND D. A. BUXTON

Chemical Defence Establishment, Porton Down, Salisbury, Wiltshire (Great Britain)

The behavioural effects of anticholinergic drugs have long been recognised. The typical syndrome produced by naturally occurring alkaloids like atropine and hyoscine, with hallucinations, delirium and confusion, was described in detail by De Boor in 1956. The psychotomimetic properties of some piperidyl benzilates, a new group of synthetic anticholinergic agents, were reported by Abood *et al.* in 1958. Subsequently Abood *et al.* (1959) claimed that there was no correlation between the anticholinergic potency and the psychotomimetic activity in man of a series of these piperidyl benzilates. The implication of this finding is, presumably, that the drugs are not acting as competitive antagonists of acetylcholine at central cholinoceptive sites to produce their behavioural effects but that other mechanisms are involved. It should be noted however that Abood *et al.* attempted to correlate psychotomimetic activity with anticholinergic potency measured on peripheral tissues; the latter may have little relevance to central potency of the drugs.

The other main piece of evidence which has been used to cast doubt on the view that these drugs produce their behavioural effects by a central anticholinergic mechanism is that these effects are apparently not readily reversed by anticholinesterase agents (Abood *et al.*, 1959; Gershon and Olariu, 1960; Bell *et al.*, 1964).

Against this there are reports that the coma produced by high doses of atropine when used in the treatment of mental disease, can be reversed by the anticholinesterase drug physostigmine (Forrer and Miller, 1958; Dolmierski and Smoczynski, 1963), and Albanus (1970) showed that the behavioural effects of atropine in dogs were readily reversed by a number of anticholinesterase agents.

The results of other animal experiments have also tended to suggest that central cholinergic mechanisms are involved in the behavioural effects of these drugs. For example, Herz (1963) showed a parallelism between the effects of a number of anticholinergic drugs upon the electrical activity of the brain and on the behaviour of animals. He interpreted this result as indicating that a common cholinergic mechanism was involved in the two actions. White and Carlton (1963) studied the effects of some piperidyl benzilates on the electrical activity of the brain of rabbits and also expressed the view that their results supported the hypothesis that these psychotomimetic benzilates acted via a central anticholinergic mechanism.

One possible method of resolving this controversy would seem to be to determine whether a correlation exists between the central, rather than the peripheral, anti-

References pp. 124–125

cholinergic potencies of a series of drugs and their potencies in producing behavioural changes. There are two difficulties however, in this approach. One is to find a satisfactory test method for determining central anticholinergic activity and the other is to select suitable aspects of behaviour for study.

The drug oxotremorine seems to lend itself to a test procedure that satisfies the first requirement. This is a muscarinic agonist which produces marked muscular tremors in a variety of species of animals. These tremors appear to result from stimulation of central muscarinic receptors (Bebbington and Brimblecombe, 1965) so that the potency of drugs in blocking the tremors gives a measure of their anticholinergic potency. It should be noted, however, that there has been very considerable discussion as to the central mode of action of oxotremorine especially since many authors have found that it produces elevations in brain acetylcholine levels (Holmstedt and Lundgren, 1966). This raises the possibility that its action may be an indirect one mediated by acetylcholine. However, since its effects are not potentiated by anticholinesterase agents and it has no nicotinic actions and, contrary to previous reports, the time course of the tremors does not seem to parallel the time course of the elevation in acetylcholine levels (Cox and Potkonjak, 1969) the conclusion that its tremorogenic actions result from direct interaction with central muscarinic receptors still seems a valid one.

Brimblecombe and Green (1967, 1968) measured the central and peripheral anticholinergic activities of 8 drugs noting, among other observations, their abilities to block the tremors and salivation induced in mice by oxotremorine. They were able to show a highly significant degree of correlation between the potency of the drugs in blocking oxotremorine-induced tremors and their potency in elevating the electrocortical arousal threshold in cat *encéphale isolé* preparations. This latter is a phenomenon connected with the well-known property of these drugs to produce an apparent dissociation between behaviour and electrocortical activity. The finding provided quantitative evidence for a cholinergic link in the electrocortical arousal system or the mesodiencephalic activating system which is interrupted by anticholinergic drugs. The significance of this interruption in terms of the behaviour of the whole animal is not clear, however.

The study reported here used the same 8 anticholinergic drugs that were employed by Brimblecombe and Green (1967, 1968). Their effects on 3 aspects of rat behaviour were studied. The types of behaviour chosen were as varied as possible and included spontaneous motor activity, emotional and conditioned behaviour.

METHODS

a. Measurement of anticholinergic potency

The method used has been described by Brimblecombe and Green (1968). Essentially, mice were injected intraperitoneally with the anticholinergic drug 15 min before the intravenous injection of 100 μg/kg oxotremorine. The animals were observed for the presence of salivation and/or tremors. An ED_{50} for block of salivation and of tremors was calculated using Thompson's (1947) method of moving averages employing the tables calculated by Weil (1952).

b. Behavioural test methods

(i) Measurement of spontaneous activity
Photocell activity cages, 56 cm × 24 cm × 16.5 cm were used. A beam of infrared light crossed the short axis of the cage and one rat per cage was used. Interruptions of the beam were recorded on decatron counters.

The experimental procedure was that rats, in groups of 12, were injected subcutaneously with 25 mg/kg of the drug or with an equivalent volume of physiological saline which was used as the solvent. Recordings began 15 min later and were made for a period of 2 h. Activity scores were summed for each group and mean scores were compared with controls for significant differences using Student's t-test. If the initial dose had a significant effect on activity then the experiment was repeated with half the initial dose and this procedure was continued until a dose was reached which produced no effect. Thus an approximate minimal effective dose (MED) for change in spontaneous activity was obtained for each drug.

(ii) Conditioned avoidance response tests
The conditioned avoidance response experiments were carried out in an automatic shuttlebox apparatus (Campden Instruments, London) in which 6 sec of tone (conditioned stimulus) were followed by 6 sec of scrambled shock (unconditioned stimulus). There were 2 trials/min. All experiments consisted of two training sessions of 50 trials each, performed at an interval of 24 h. All the anticholinergic drugs were first tested at a dose of 25 mg/kg injected subcutaneously 15 min before the start of the first training session. No treatment preceded the second session. Thus the first session tested the effects of the drug on the acquisition of the response and the second session tested for retention of the response and examined continued learning in the non-drugged state. At each dose 10 drug-treated and 10 control rats were used. Mean scores were calculated for avoidance respondings and intertrial crossings in each session and were compared with corresponding control scores for significant differences using Student's t-test.

(iii) Open-field test
The procedure used was that described by Brimblecombe (1963) except that the open-field apparatus used was as described by Broadhurst (1957).

Essentially, groups of 8 rats were injected subcutaneously with the anticholinergic drug or an equivalent volume of physiological saline and placed singly in the open field for a period of 3 min at a time either 1.5 or 3 h after injection. The animals were then observed for number of times rearing, preening and defaecating, number of squares traversed, number of faecal boluses passed and number of squares traversed away from the periphery of the field. Each of these parameters in the drug-treated groups were compared with corresponding values in the control group for

References pp. 124–125

TABLE I

LIST OF DRUGS USED AND THEIR POTENCIES IN ANTAGONISING OXOTREMORINE-INDUCED SALIVATION AND TREMORS IN MICE

Drug no.	*Formula and name*	*ED50 (μmoles/kg) with 95% limits for antagonism of oxotremorine induced:*	
		Salivation	*Tremors*
1.	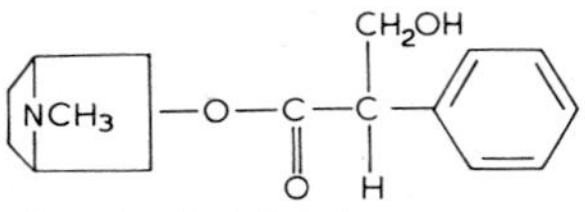 Atropine (sulphate)	0.44(0.30–0.66)	16.2(10.0–26.6)
2.	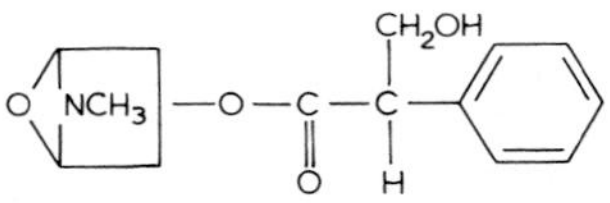 Hyoscine (hydrobromide)	0.05(0.02–0.08)	1.1 (0.60– 2.5)
3.	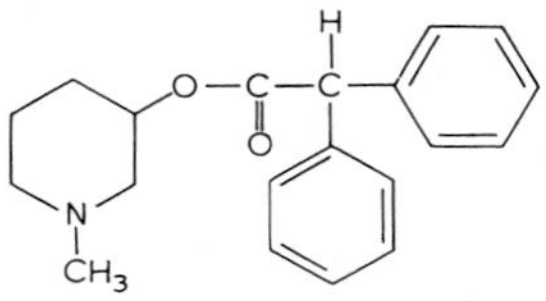 N-methyl-3-piperidyl benzilate (HCl)	1.05(0.53– 1.9)	1.8(0.96– 3.2)
4.	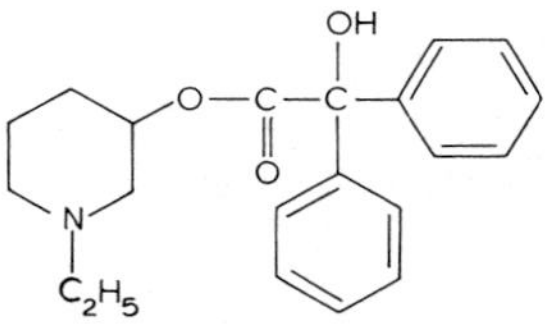 N-ethyl-3-piperidyl benzilate(HCl)	7.1(4.1 –12.1)	2.7(1.4 – 5.3)
5.	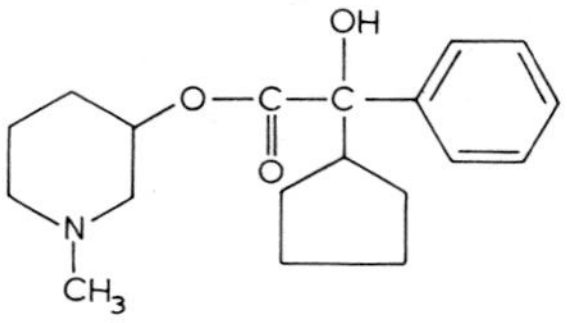 N-methyl-3-piperidyl phenylcyclopenty glycollate (HCl)	1.2(0.38– 1.6)	2.0(1.5 – 2.7)
6.	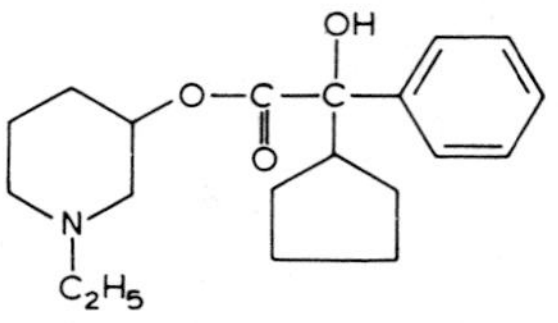 N-ethyl-3-piperidyl phenylcyclopentyl glycollate (HCl)	1.5 (1.2 –2.0)	3.1 (1.8 – 5.3)

TABLE I *(Continued)*

Drug no.	*Formula and name*	*ED 50 (μmoles/kg) with 95% limits for antagonism of oxotremorine induced:*	
		Salivation	*Tremors*
7.	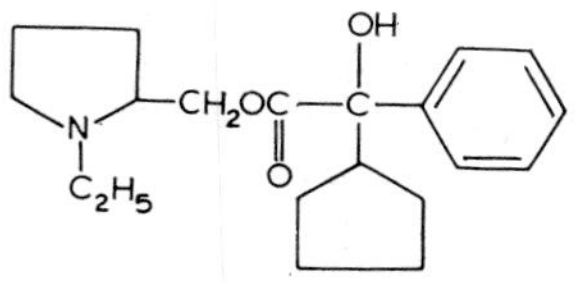 N-ethyl-2-pyrrolidylmethylphenylcyclopentyl glycollate, PMCG (HCl)	1.7 (0.4 –7.1)	3.5 (2.2 – 5.7)
8.	70% of 7 + 30% of 6, Ditran	0.77(0.15–1.2)	0.88(0.50 –1.4)

significant differences using Student's *t*-test. Testing of each drug commenced at 10 mg/kg and progressively lower doses were used (reduced by half in each case) until a no effect level was reached. Thus an approximate minimal effective dose (MED) was obtained for each drug.

c. Drugs used

The drugs used are listed in Table I. Atropine sulphate was purchased from B.D.H. Ltd and hyoscine hydrobromide from Macfarlan Smith Ltd. All the other drugs were prepared in these laboratories and were used as the hydrochlorides. Doses were given in mg/kg as indicated above but for calculation of correlation coefficients these were converted into μmoles/kg.

RESULTS

The potencies of the drugs in blocking oxotremorine-induced salivation and tremors are also given in Table I. These results are taken from Brimblecombe and Green (1968).

Results from the spontaneous activity experiments are given in Table II. The minimal effective doses for increases in spontaneous activity are given for all 8 drugs. Coefficients of correlation were calculated for these values with potencies of the drugs in blocking oxotremorine-induced salivation and tremors. The coefficients are given in Fig. 2.

All 8 drugs caused significant increases in the numbers of intertrial crossings in the shuttle box during the first training session. Three drugs (Nos. 3, 4 and 6) produced significant increases in avoidance scores during the first training session and 3 drugs (Nos. 3, 6 and 8) produced significantly fewer avoidances than controls during the second training session. The doses required to produce these effects are given in Table III. Two drugs (Nos. 3 and 6) produced both these effects, and results

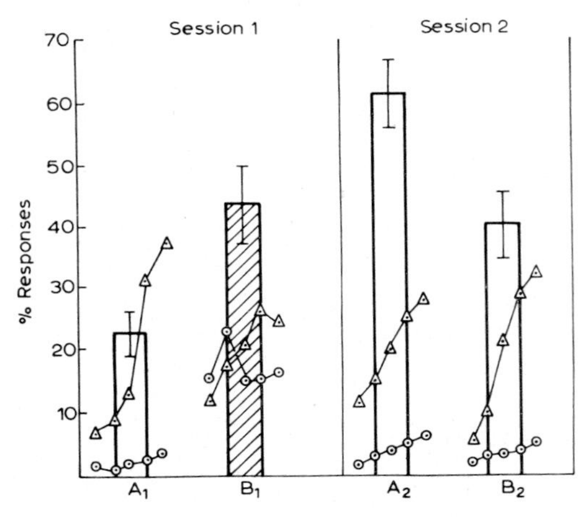

Fig. 1. Effects on conditioned avoidance (shuttlebox) of compound 6 (25 mg/kg). Vertical blocks: mean percentage avoidance ± S.E.M. A_1 and A_2, control group; B_1 and B_2, drug group, injected 15 min before first session (shaded block). (△) percentage avoidances occurring during the 5 successive 10-trial periods of the daily session; (⊙) percentage intertrial responses occurring during the 5 successive 10-trial periods of the session.

	6	5	4	3	2	1
(1) SQUARES CROSSED (open-field)	0.89 (two underlinings)	0.21	0.29	0.43	0.85 (two underlinings)	
(2) DEFAECATIONS (open-field)	0.65	0.49	0.66	0.26		
(3) SPONT. ACTIVITY	0.60	0.29	0.77 (one underlining)			
(4) INTERTRIAL CROSSES (shuttlebox)	0.04	0.75 (one underlining)				
(5) BLOCK OF TREMORS (after oxotremorine)	0.48					
(6) BLOCK OF SALIVATION (after oxotremorine)						

df = 6 in all cases except correlations with "Defaecations" where *df* = 5

One underlining shows $P < 0.05$

Two underlinings show $P < 0.01$

Fig. 2. Correlation coefficients.

of an experiment using compound No. 6 are shown in Fig. 1. Coefficients of correlation were calculated between MEDs for increases in intertrial crossings and ED_{50}s for blockade of oxotremorine-induced salivation and tremors and MEDs for increases in spontaneous activity. (Fig. 2).

In the open-field test all 8 drugs caused significant increases in the number of squares traversed, *i.e.*, in the total degree of ambulation, and all except compound No. 5 caused significant decreases in the amount or frequency of defaecation. The minimal effective doses for these effects are shown in Table IV, and various coefficients of correlation were calculated using these values which are shown in Fig. 2.

TABLE II

MINIMAL EFFECTIVE DOSES (MED) OF THE DRUGS (mg/kg) REQUIRED TO PRODUCE INCREASES IN SPONTANEOUS ACTIVITY OF RATS

Drug no.	*MED*
1	3.13
2	0.19
3	0.39
4	1.56
5	0.78
6	0.78
7	0.39
8	1.56

DISCUSSION

All the drugs studied produced marked increases in spontaneous motor activity but their potency in this respect was not correlated with their potency in blocking oxotremorine-induced salivation or tremors. A similar lack of correlation with peripheral anticholinergic activity has been reported by Harris (1961) and Abood (1968).

The results of the conditioned avoidance response experiment are complex. Three out of the 8 drugs used (Nos. 3, 4, and 6) caused significant increases in avoidances as compared with the controls in the first session and 3 out of the 8 (Nos. 3, 6 and 8) caused significant decreases in the second session. All the drugs produced significantly more intertrial crossings than the controls in the first session, an effect statistically correlated with their potency in increasing spontaneous activity. It is tempting to conclude that these increases in avoidances result merely from hyperactivity, the existence of which is indicated by the large number of intertrial crossings. Animals which are crossing the barrier frequently may be expected to cross randomly during conditioned stimulus periods or intertrial periods. However, examination of the data in Table III shows that the drugs which are most potent in producing hyperactivity

TABLE III

SUMMARY OF SHUTTLEBOX RESULTS

Drug no.	*MED (mg/kg) for increase in intertrial crossings Session 1*	*MED (mg/kg) for increase in avoidances Session 1*	*MED (mg/kg) for decrease in avoidances Session 2*
1	25.0	—	—
2	12.5	—	—
3	0.8	0.8	6.3
4	12.5	25.0	—
5	12.5	—	—
6	1.6	3.1	1.6
7	12.5	—	—
8	12.5	—	25.0

References pp. 124–125

TABLE IV

MINIMAL EFFECTIVE DOSES FOR DECREASES IN DEFAECATION AND INCREASES IN AMBULATION IN THE OPEN FIELD

Drug	*MED (mg/kg) for decrease in defaecations*	*MED (mg/kg) for increase in square count*
1 Atropine	0.80	1.60
2 Hyoscine	0.20	0.05
3 N-Methyl 3 piperidyl benzilate	0.10	0.10
4 N-Ethyl 3 piperidyl benzilate	1.00	5.00
5 N-Methyl 3 piperidyl phenyl cyclopentyl glycollate	No effect	0.05
6 N-Ethyl 3 piperidyl phenyl cyclopentyl glycollate	0.10	0.20
7 P.M.C.G.	0.50	0.20
8 Ditran	0.10	0.10

are not necessarily the ones which produce increased avoidance. Additionally, a check of the probability of the occurrence of a cross during the conditioned stimulus and intertrial period was made. If an animal is crossing the barrier randomly, *i.e.*, without regard to the stimuli present, the probability of his crossing during an intertrial period (18 sec) and during the conditioned stimulus period (6 sec) should be in the ratio of 3:1 (assuming that the animal will always cross during the shock period). In fact checks on the data showed that this ratio was less than 3:1 indicating a learned response to the conditioned stimulus. Reference to Fig. 1 also indicates that whereas there was a progressive increase in the number of avoidances made by the drugged animals throughout session 1, the number of intertrial crossings reached a maximum part way through the session and then fell away.

The conclusion to be drawn from this evidence seems to be that all the increased avoidance shown by the anticholinergic drug-treated animals did not result directly from random movements which happened to occur at the appropriate time. However, the hyperactivity of these animals may have provided the opportunity for them to make "avoidance" responses earlier than controls and by virtue of this fact they may have been conditioned more quickly. This is not a complete explanation, however, since, as was pointed out earlier, the drugs which were most potent in producing hyperactivity were not necessarily those which were most potent in producing increased avoidance. The findings are in general agreement with those of Oliverio (1967) who reported that hyoscine enhanced the performance of naive mice in a conditioned avoidance response situation.

There are several possible explanations of the fact that 3 drugs produced significantly fewer avoidances than controls during the second session in the shuttlebox. Acquisition of the conditioned response was apparently not affected since, as has already been discussed, some conditioning did occur during the first session. This might imply that processes of consolidation or of recall were being affected by the drugs or alternatively that there was simply a lack of transfer of conditioning from the drugged to the undrugged state, *i.e.*, that this was drug-dependent or state-dependent learning. Such an effect of hyoscine has been postulated by Evans and

Patton (1970) who used rats in a one-trial conditioned suppression technique. The data do not permit any conclusions to be drawn concerning the relative likelihood of these possibilities.

All the compounds studied produced an increase in the amount of ambulation in the open field. Reference to Fig. 2 reveals that the potency of drugs in this respect was not correlated with their potency in increasing spontaneous activity in photocell cages, *i.e.*, it would appear that this increase in open-field ambulation is not precisely analogous to a simple increase in spontaneous activity but that the stress of the open-field situation is modifying this response to the drug. Anderson (1938) reported a similar lack of correlation.

All but one of the compounds (No. 5) produced a decrease in the amount or frequency of defaecation in the open field. This effect would be expected on pharmacological grounds in that all these drugs, being potent anticholinergics, will have spasmolytic actions on the gut. However, the fact that this effect is highly significantly correlated with the increase in the open-field ambulation suggests that it might have some behavioural significance, possibly in reducing emotionality since open-field defaecation is generally agreed to be an emotional response.

The basic object of this study was to determine whether the activities of any of these anticholinergic drugs in modifying various aspects of behaviour were correlated with their anticholinergic activities as measured by ability to antagonise oxotremorine-induced salivation or tremors in mice. In the event only two significant correlations emerged. The highly significant correlation between blockade of oxotremorine-induced salivation (a peripheral effect) and increase in open-field ambulation (presumably a central effect) is surprising and no explanation can be offered for it. The other correlation, between blockade of oxotremorine tremors and intertrial crossings in the shuttlebox, suggests that the latter effect is mediated through cholinergic mechanisms. There are several possible reasons for the lack of correlation between effects of the drugs on the other aspects of behaviour and their potencies in antagonising oxotremorine effects. The most obvious explanation is that cholinergic systems are not involved in these aspects of behaviour but it would be premature and unjustified to draw such a conclusion at this stage. Firstly, it seems reasonable to suppose that a complex pattern of behaviour might be controlled by physiological systems involving more than one neurotransmitter substance and that correlations with pharmacological measures of purely anticholinergic activity may therefore be unlikely. Secondly, the variability inherent in most measures of behaviour is high and therefore it may be necessary to use more drugs than were included in the present study in order to demonstrate any correlations which may exist. Thirdly, all the behaviour studies were carried out in rats although mice were used for the oxotremorine experiments. Further studies are planned in which these factors will be taken into account.

In the meantime, if the results are considered qualitatively rather than quantitatively, they do not conflict with the findings of many other workers which tends to suggest that anticholinergic drugs block inhibitory cholinergic systems thereby producing a state of activation.

References pp. 124–125

SUMMARY

1. The effects of 8 anticholinergic drugs were studied on 3 aspects of rat behaviour. These included spontaneous motor activity, open-field behaviour and avoidance conditioning in a shuttlebox.

2. The peripheral and central anticholinergic activities of the 8 drugs were also measured in terms of their potencies in antagonising oxotremorine-induced salivation and tremors respectively in mice.

3. All the drugs produced marked increases in spontaneous motor activity but their potencies in this respect were not correlated with their anticholinergic activity.

4. All the drugs caused significant increases in intertrial crossings in the shuttlebox and this may have in part, but not completely, contributed to the facilitation of avoidance conditioning produced by 3 of the drugs. Three drugs produced a decrement in performance in trials carried out 24 h after drug administration.

5. In the open-field test all the drugs caused significant increases in ambulation and all except one compound caused significant decreases in the amount or frequency of defaecation. The minimal effective doses of the drugs in producing these two effects on open-field behaviour were correlated with each other but not with measures of central anticholinergic potency.

6. The results suggest that these anticholinergic drugs block inhibitory cholinergic systems thereby producing a state of activation. It was not possible, however, to demonstrate this quantitatively and the possible reasons are discussed.

ACKNOWLEDGEMENTS

The authors acknowledge the valuable technical assistance by Mrs. Patricia Muir and Mrs. Maureen Saunders.

REFERENCES

ABOOD, J. G. (1968) The psychotomimetic glycollate esters. In *Drugs Affecting the Central Nervous System*, A. BURGER (Ed.), Marcel Dekker, New York, pp. 127–167.

ABOOD, L. G., OSTFELD, A. M. AND BIEL, J. H. (1958) A new group of psychotomimetic agents. *Proc. Soc. exp. Biol. (N.Y.)*, **97**, 483–486.

ABOOD, L. G., OSTFELD, A. M. AND BIEL, J. H. (1959) Structure–activity relationships of 3-piperidyl benzilates with psychotogenic properties. *Arch. int. Pharmacodyn.*, **120**, 186–200.

ALBANUS, L. (1970) Studies on central and peripheral effects of anticholinergic drugs. *Forsvarets Forskningsanstalt Reports Vol. 4, No. 4*, 1–17.

ANDERSON, E. E. (1938) The interrelationship of drives in the male albino rat. II. Intercorrelations between 47 measures of drives and of learning. *Comp. Psychol. Monogr.*, **14**, 1–119.

BEBBINGTON, A. AND BRIMBLECOMBE, R. W. (1965) Muscarinic receptors in the peripheral and central nervous system. *Adv. Drug. Res.*, **2**, 143–172.

BELL, C., GERSHON, S., CARROLL, B. AND HOLAN, G. (1964) Behavioural antagonism to a new psychotomimetic: JB-329. *Arch. int. Pharmacodyn.*, **147**, 9–25.

BRIMBLECOMBE, R. W. (1963) Effects of psychotropic drugs on open-field behaviour in rats. *Psychopharmacologia (Berl.)*, **4**, 139–147.

BRIMBLECOMBE, R. W. AND GREEN, D. M. (1967) Further evidence for cholinergic synapses in the mesodiencephalic activating system. *J. Physiol. (Lond.)*, **194**, 16–17.

BRIMBLECOMBE, R. W. AND GREEN, D. M. (1968) The peripheral and central actions of some anticholinergic substances. *Int. J. Neuropharmacol.*, **7**, 15–21.
BROADHURST, P. L. (1957) Determinants of emotionality in the rat. I. Situational factors. *Brit. J. Psychol.*, **48**, 1–12.
COX, B. AND POTKONJAK, D. (1969) The relationship between tremor and change in brain acetylcholine concentration produced by injection of tremorine or oxotremorine in the rat. *Brit. J. Pharmacol.*, **35**, 295–303.
DE BOOR, W. (1956) *Pharmakopsychologie und Psychopathologie* Springer, Berlin.
DOLMIERSKI, R. AND SMOCZYNSKI, S. (1963) Studies on the therapeutic value of atropine coma. *Bull. pol. med. Hist. Sci.*, **2**, 64–67.
EVANS, H. L. AND PATTON, R. A. (1970) Scopolamine effects on conditioned suppression. Influence of diurnal cycle and transitions between normal and drugged states. *Psychopharmacologia (Berl.)*, **17**, 1–13.
FORRER, G. R. AND MILLER, J. J. (1958) Atropine coma: somatic therapy in psychiatry. *Amer. J. Psychiat.*, **115**, 455–458.
GERSHON, S. AND OLARIU, J. (1960) JB 329—A new psychotomimetic. Its antagonism by tetrahydroaminoacrin and its comparison with LSD, mescaline and sernyl. *J. Neuropsychiat.*, **1**, 283–292.
HARRIS, L. S. (1961) The effect of various anticholinergics on spontaneous activity of mice. *Fed. Proc.*, **20**, 395.
HERZ, A. (1963) Excitation and inhibition of cholinoceptive brain structures and its relationship to pharmacological induced behaviour changes. *Int. J. Neuropharmacol.*, **6**, 133–142.
HOLMSTEDT, B. AND LUNDGREN, G. (1966) Tremorogenic agents and brain acetylcholine. In *Mechanisms of Release of Biogenic Amines*, U.S. VON EULER, S. ROSELL AND B. UNVAS (Eds.), Pergamon Press, Oxford, pp. 439–468.
OLIVERIO, A. (1967) Contrasting effects of scopolamine on mice trained simultaneously with two different schedules of avoidance conditioning. *Psychopharmacologia (Berl.)*, **11**, 39–51.
THOMPSON, W. R. (1947) Use of moving averages and interpolation to estimate median-effective dose. *Bact. Rev.*, **11**, 115–145.
WEIL, C. S. (1952) Tables for convenient calculation of median-effective dose (LD50 or ED50) and instructions in their use. *Biometrics*, **8**, 249–263.
WHITE, R. P. AND CARLTON, R. A. (1963) Evidence indicating central atropine-like actions of psychotogenic piperidyl benzilates. *Psychopharmacologia (Berl.)*, **4**, 459–471.

DISCUSSION

RICK: The correlation you showed between defaecation and ambulation was a positive one but I got the impression that the rate of defaecation was going down as that of ambulation was going up.

BRIMBLECOMBE: The correlation was a positive one in that we correlated minimal effective dose for increase in ambulation with minimal effective dose for decrease in defaecation, but the effect was as you state.

RICK: In the open field, ambulation on day 1 is now taken to be of a complex character, partly exploratory and partly escape behaviour (Whimbey and Denenberg, 1967). It might become clearer if you did an experiment on day 2 because there ambulation is more clearly exploratory in nature.

BRIMBLECOMBE: This may well be right and may explain why we did not get a correlation between open-field ambulation and our other two measures of hyperactivity.

COX: I would like to refer to your slide which reports the effects of N-ethyl-3 piperidyl benzilate on shuttlebox avoidance. The histograms represent, I presume, the group means and standard error or deviation for these groups. The statistical significance of the differences between the control and drug groups has not been mentioned. It appears that there may be a slight difference between the control and drug scores on the percentage avoidance index, while there are obviously significant differences between these scores and the intertrial crossing index. The drug therefore appears to increase the number of intertrial crossings but to lack a significant effect on the actual avoidance. Furthermore, the drug only appears to affect intertrial crossings on the session immediately following its administration. This would be consistent with our studies in Nottingham which tend to indicate that in considering behavioural measures of "learning", ACh-related agents affect some general component of performance rather than a component of the learning process *per se*.

BRIMBLECOMBE: The histograms represent group means and standard errors. The drug-treated animals showed a significantly greater number of avoidances than controls on the first day and a significantly lower number on the second day. As I explained, we believe that the effect on avoidance on the first day cannot be explained solely on the basis of increased activity, or, as you put it, on some general component of performance.

KING: There are some data rather different from yours which show that under atropine the rat takes longer to start to learn an experimental problem, but once it has started to learn, the rate at which it learns is not different from that of control animals (Warburton, 1969). I wondered if in your experiments there was a point when the number of intertrial crossings started to return to control level and at this point real learning started to take place in the context of the shuttlebox avoidance situation.

BRIMBLECOMBE: Yes, it appears that during the early part of the first session the increase in avoidance may be due to hyperactivity but during the latter part of the session I think that true conditioning or learning is taking place, because, as you say, the number of intertrial crossings started to return to control level.

REFERENCES

WARBURTON, D. M. (1969) Behavioural effects of central and peripheral changes in acetylcholine systems. *J. comp. physiol. Psychol.*, **68**, 56–64.

WHIMBEY A. E. AND DENENBERG V. H. (1967) Two independent behavioural dimensions in open-field performance. *J. comp. physiol. Psychol.*, **63**, 500–504.

Centrally Acting Cholinolytics and the Choice Behaviour of the Rat

A. M. VAN DER POEL

Department of Fundamental Pharmacology, University of Leiden, Wassenaarseweg 62 Leiden (The Netherlands)

Rats typically alternate their choice directions (left–right) during successive trials in a T-maze. One of the known effects of centrally acting, cholinolytic drugs is their suppressing influence on this alternation behaviour (Meyers and Domino, 1964; Parkes, 1965; Douglas and Isaacson, 1966; Squire, 1969). Apparently it is a specific effect, because the doses required are low, and representatives of other classes of psychotropic drugs do not show such distinct effects on alternation (Sinha *et al.*, 1958; Zahner *et al.*, 1961; Grandjean and Bättig, 1962; Parkes, 1965). Dember and Fowler (1958) called attention to the phenomenon of alternation as being possibly a suitable behavioural test for studying processes of more general interest such as memory and perception. By excluding the use by the animal of external, directional clues originating from the previous trials (*e.g.*, odour trial), Douglas (1966) and Squire (1969) showed that some kind of memory has to be involved in alternation. Therefore, it seemed important to investigate whether the effect of cholinolytics on alternation behaviour could be interpreted in terms of memory and perception. The idea was to let the point of maximal effect of an appropriate cholinolytic drug coincide with each of the 3 memory stages (registration, consolidation and retrieval). In order to do so, it was necessary to lengthen the intertrial interval beyond the duration of action of such a drug, because only then could the effects on registration, consolidation and retrieval be sufficiently differentiated.

If the drug-state of an animal between the registration and the retrieval phase is changed, learning often appears to be impaired, whereas if the drug-state is kept constant, no such impairment is observed. This phenomenon has been called "state-dependent learning" and has been shown to occur after administration of cholinolytics (Overton, 1966; Oliverio, 1968). Although available data (Meyers and Domino, 1964; Parkes, 1965; Douglas and Isaacson, 1966) suggest that "state-dependent learning" does not play a role in the effect of cholinolytics on alternation behaviour if short intertrial intervals are used, it was decided to look for this phenomenon in a separate experiment with a long intertrial interval by maintaining the animals under the influence of a cholinolytic drug during all trials.

Effects on motivational factors, which could conceivably affect alternation behaviour, have been considered by analyzing the behaviour of treated and untreated

References pp. 136–137

Explore (sniffing with scanning movements of the head, mostly accompanied by locomotion), Rearing (standing up on the hindlegs), and Intention movement (shown at the choice point, in its lowest intensity: a slight movement of the head, directed to one of the arms; in its highest intensity: the animal entered an arm with head and forepaws and subsequently retreated).

Experimental design — one PCMG treatment

Three experiments were performed. In each experiment 4 groups of 10 trained animals were used. According to a 4 $\times$ 4 Latin-square scheme each group was successively subjected to each of 4 possible procedures. As shown in Fig. 2, column 2, each procedure consisted of 3 trials (indicated by I, II and III), separated by a first intertrial interval of 2 h and a second one of only 1 min. Three injections were given: the first 20 min before trial I, the second immediately after trial I and the third 20 min before trial II. One of the injections of procedures B, C and D contained 5 mg/kg PCMG, while all other injections contained the vehicle (phosphate–citric acid buffer, 10 times diluted, pH 6). Dose and time were chosen on the basis of preliminary experiments. The same animals were subsequently used in a second experiment.

Two PCMG treatments

Eighty trained animals, previously used in the first series of experiments ("one PCMG treatment"), were administered 2.5 mg/kg PCMG twice: 20 min before trial I and 20 min before trial II. There were 2 trials with an intertrial interval of 2 h on 1 day only. Three animals refused to run within 5 min and were therefore discarded. This experiment was meant to maintain the drug-state at the same level during each of the two trials.

Scopolamine and actual running speeds

The remaining 40 trained animals were subsequently divided into two groups of 20 animals each. Actual running speeds in the first 40 cm of the leg and the first 40 cm of the arms of the T-maze were measured by recording the time to cover these distances. There were two trials a day, separated by an intertrial interval of 15 min, for two days. Because in this case there were no special reasons for using PCMG and in view of the small amount of this drug available, scopolamine was used. Fifteen minutes before the first trial 1 mg/kg scopolamine or an equivalent amount of saline was administered. A simple crossover design was used.

Drugs used

PCMG as its chloride (Abbott Laboratories), and scopolamine-HBr (Sandoz) were used. Solutions were freshly prepared daily. PCMG was injected in a solution containing (mM): citric acid 3.7; Na_2HPO_4 12.6. Scopolamine was dissolved in saline.

All injections were given intraperitoneally. Successive injection days were separated by at least one day of rest.

Statistical evaluation

Latin-square analyses were performed according to the method described by Natrella (1963). To derive two-tailed probability level (P) for different ranges (w) the following procedure was used: for 17 probability levels 17 values of ω were obtained from the tables provided by Harter (1960a, b) and Owen (1962), each for 4 groups and 6 degrees of freedom. It was possible to get good fit by means of a fifth degree polynomial for log P against w, which was used to calculate the P's belonging to the observed w-values. These were combined according to De Jonge (1964).

The average chance level of alternation was calculated by means of the "empirical chance" method, described by Douglas and Isaacson (1965). To obtain this value the choice directions of the first trial of successive days during the final stage of the training period were compared. Alternation percentages observed under different conditions, were compared by means of the Chi-square test for two independent samples (Siegel, 1956). Douglas and Isaacson (1965) and Squire (1969) have shown that in the case of alternation it is permitted to assume that repeated samples taken from the same group of animals are independent.

Yates' method for obtaining estimates of main effects and interactions was applied to the measurements of the actual running speeds (Yates, 1937). There were 3 factors, each at two levels: leg *vs* arm; trial I *vs* trial II; drug *vs* non-drug.

RESULTS

During the final phase of the training (last 5 days) the animals alternated at an average rate of 70%, whereas the average chance level of alternation was 44.7%.

One PCMG treatment

As shown in Fig. 2, column 3, treatment before trial I and before trial II had a marked, negative effect on alternation behaviour, whereas treatment immediately after trial I had much less effect. In fact, in two out of three experiments there was no effect at all, but the third experiment produced an extremely low figure, which, however, could not be reproduced with the same animals later.

The fourth column of Fig. 2 shows that administration of PCMG 140 min, and 120 min before trial II (procedures B and C, respectively) had no effect on the alternation percentage, obtained by comparison of choice directions shown during trial II and trial III. Administration of PCMG 20 min before trial II (procedure D) produced only a moderate decrease of this alternation percentage.

Fig. 3 shows the median values of latencies, choice times and running times obtained by averaging the results of trial I of the 3 experiments. Under the influence of

References pp. 136–137

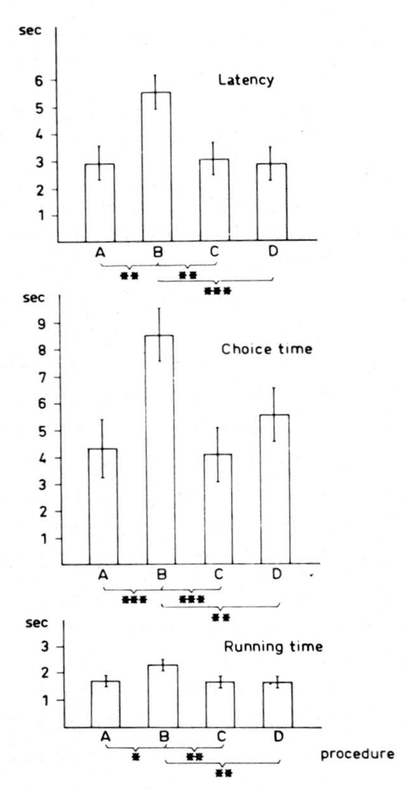

Fig. 3. The influence of administration of 5 mg/kg PCMG i.p. on median times of trial I (± S.E.M., calculated on the basis of Latin-square analysis). Coding of procedures as in Fig. 2: A, Control; B, PCMG 20 min before trial I; C, PCMG immediately following trial I; D, PCMG 20 min before trial II. Significant differences are indicated by braces and asterisks: *, $P < 0.05$; **, $P < 0.01$; ***, $P < 0.001$. For the definitions of latency, choice time and running time see text.

the drug (procedure B) they all appear to be significantly prolonged: the latency and the choice time are about twice as long as in the control situations (procedures A, C and D), whereas the increase of the running time is much smaller.

In Fig. 4, the median times of trial II have been collected. Here, too, administration of PCMG 20 min before (procedure D) leads to a 2-fold, significant increase in latency and choice time, whereas the increase in the running time is again much smaller. One hundred and forty minutes and 120 min after administration the effect of PCMG on the median times had not yet vanished: all values observed with procedures B and C were intermediate between those found with procedure A (controls) and procedure D (20 min after administration). The choice times observed 140 min and 120 min after administration and the running times 120 min after administration are significantly different from control.

There are only slight differences between the median times of the saline-treated animals observed during trial I (procedures A, C and D, Fig. 3) and trial II (procedure A, Fig. 4), indicating a stable baseline performance of control animals.

An analysis of the behavioural elements shown in different parts of the T-maze demonstrated significant increases in the frequency of Explore and Intention movements under the influence of 5 mg/kg PCMG. The following significant increases in the average frequency per animal were noted: during the latency of trial I Explore rose from 0.9 to 1.5 ($P < 0.05$); at the choice point during both trial I and trial II Explore increased from 0.2 to 2.3 ($P < 0.01$) and from 0.3 to 1.2 ($P < 0.01$), respectively; Intention movements rose from 1.1 to 5.1 ($P < 0.05$) during trial I and from 1.0 to 3.1 ($P < 0.05$) during trial II. Although not finding expression in the data collected there was also an obvious increase in the intensity of the Intention movements shown by PCMG-treated animals. The frequency of Rearing showed no increase. There were no systematic differences between groups and injection days in any of the criteria tested.

Two PCMG treatments

The alternation percentage observed under this condition was 40.3%, which is not significantly different from the previously determined chance level of alternation ($X^2 = 0.42$, $df = 1$).

Scopolamine and actual running speeds

Table I summarizes the results. A factorial analysis, performed on the median times per group of 20 animals, revealed a significant influence of the part of the maze (g = −7.25; w = 3.9), indicating faster running in the arm of the T-maze. There was

TABLE I

ALTERNATION PERCENTAGE AND MEDIAN ACTUAL RUNNING-SPEEDS (cm/sec), OBSERVED IN THE FIRST 40 cm OF BOTH LEG AND ARM OF THE T-MAZE DURING TRIAL I AND II AFTER THE ADMINISTRATION OF SALINE OR 1 mg/kg SCOPOLAMINE

Treatment	*Trial no.*	*Running speeds*		*Alternation (%)*
		leg	*arm*	
Saline	I	43.2	48.5	72.5
	II	37.2	43.2	
Scopolamine	I	37.2	47.1	30.0
	II	36.4	47.1	

no influence of the trial number or the administration of scopolamine. None of the interactions was significant.

As expected, the scopolamine-treated animals alternated at a low rate (30%), which was significantly different from the alternation percentage of the saline treated group (72.5%; $X^2 = 14.46$, $df = 1$), but did not reliably differ from the previously determined chance level of alternation (44.7%; $X^2 = 2.79$, $df = 1$).

References pp. 136–137

DISCUSSION

This study has made it clear that PCMG has an effect on processes relevant to alternation behaviour which occur simultaneously or almost simultaneously with trial I and trial II. The effect of the administration prior to trial I, and the absence of an effect of the administration immediately following trial I indicates that PCMG affects the registration and not the consolidation of information, which rats require to perform a correct alternation response. The effect of the administration prior to trial II is less easy to interpret. During trial II stored information is recalled and compared with newly entered information. Consequently, during trial II there is registration as well as retrieval. Therefore, an effect of PCMG on retrieval could not be demonstrated unambiguously.

These conclusions are in agreement with the results obtained by Bureš *et al.* (1964). Studying the passive avoidance behaviour of rats, it was demonstrated that 6 mg/kg atropine sulfate did not influence consolidation and permanent memory, but significantly affected learning (registration) and retrieval.

Squire (1969) examined the effect of 1 mg/kg scopolamine hydrobromide administered 25 $\pm$ 10 min before trial I or trial II, with varying intertrial intervals. The animals were not trained to alternate. Hence, with long intertrial intervals low alternation percentages were observed and therefore no influence of administration prior to trial II could be detected at intertrial intervals of more than 1 h. Still, it was concluded that such administrations had no effect and that the observed effect at shorter intervals was due to an influence on consolidation.

The empirically determined chance level of alternation deviated slightly from a level of 50% which, theoretically, is to be expected if rats do not show any individual left–right preference (Koppenaal, 1962). This deviation was regularly observed in all other situations in which rats were given daily trials in the T-maze. The administration of PCMG prior to trial I or prior to trial II reduced the alternation percentage to slightly above the empirically determined chance level, whereas with both trials under the influence of a cholinolytic drug (PCMG; scopolamine) values just below this chance level were observed. This may indicate a direct influence of these drugs on the individual left–right preference of the rat in a T-maze.

The decrease in the alternation percentage observed after treatment with PCMG either before trial I or trial II cannot easily be explained by "state-dependent learning", since with both trials under the influence of PCMG or scopolamine alternation is strongly suppressed as well. Only with a short intertrial interval of 1 min no decrease to chance levels of alternation was noted (Fig. 4).

Measurement of the actual running speeds and analysis of the behavioural elements, shown by the animals in different parts of the maze, all make a motivational interpretation very unlikely.

If choice times and running times of control rats are compared (Figs. 3 and 4) it looks as though there is some sort of an approach gradient (*i.e.*, the animals are running faster, the nearer they approach the goal). The distances the animals have to cover are equal (50 cm), whereas the choice time is more than twice as long as the

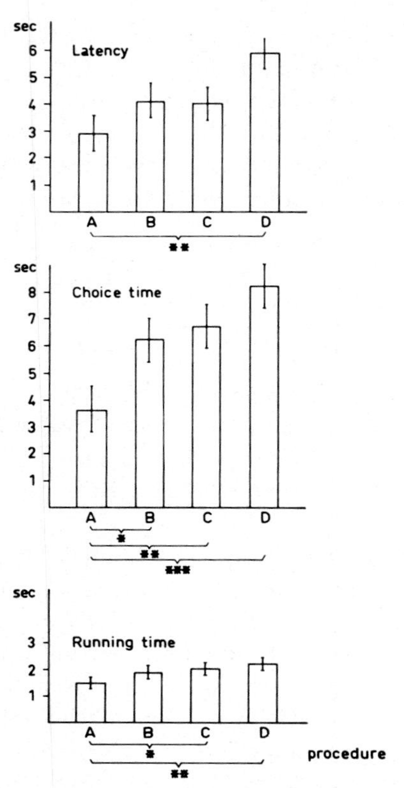

Fig. 4. The influence of administration of 5 mg/kg PCMG i.p. on median times of trial II ($\pm$ S.E.M., calculated on the basis of Latin-square analysis). Coding of procedures as in Fig. 2: A, Control; B, PCMG 20 min before trial I; C, PCMG immediately following trial I; D, PCMG 20 min before trial II. Significant differences are indicated by braces and asterisks; *, $P < 0.05$; **, $P < 0.01$; ***, $P < 0.001$. For the definitions of latency, choice time and running time see text.

running time. If the actual running speeds in the first 40 cm of leg and arm of the maze are compared, it appears that there is indeed a weak, but significant approach gradient. Seemingly, there is a change of the approach gradient after PCMG treatment, because then the ratio between choice time and running time increases to about 4, but no influence of scopolamine could be detected on actual running speeds, observed in leg and arm of the T. It is concluded that these drugs do not alter the motivational factors underlying the approach gradient.

Apart from the weak approach gradient the difference between choice time and running time in the control situation is mainly due to the performance of Intention movements by most animals at the choice point. The substantial increase of the choice time after PCMG is brought about by significant increases of the frequency of Explore and Intention movements in the choice point area.

These observations, coupled with the obvious increase in the intensity of the Intention movements after treatment may indicate that PCMG treated animals are at least as eager as controls to make a choice, but are one way or another incapable of

References pp. 136–137

making a quick decision. Most plausibly the impaired registration has something to do with this lengthening of the decision time.

The substantial increase of the latency after PCMG is attended by increases of the frequency of Explore in the startbox, which, however, reached significance only during trial I. This is taken as an indication that rats are already expecting the choice when they still are in the startbox. In other words, the decision time might begin in the startbox.

SUMMARY

1. The spontaneous alternation behaviour shown by rats during successive trials in a T-maze is dependent on the intertrial interval. Lengthening the intertrial interval beyond 1 h is attended by a decrease in the percentage of animals alternating.

2. A procedure is described to train rats to alternate at a rate of about 70% at an intertrial interval of 2 h.

3. The short acting, centrally active cholinolytic drug N-methyl-4-piperidyl cyclopentyl methylethynyl glycollate (PCMG) has been used to analyse the suppressing influence of cholinolytics on alternation behaviour.

4. With an intertrial interval of 2 h, administration of PCMG 20 min prior to trial I or 20 min prior to trial II reduced the alternation percentage to about 50%, whereas administration immediately following trial I had much less effect.

5. Analysis of running speeds and behaviour, shown in different parts of the maze, demonstrated that treated animals are as strongly motivated to run and to choose as untreated ones.

6. It is concluded that cholinolytic drugs exert their effect by disrupting registration and maybe also retrieval, whereas they probably have no effect on consolidation and motivation.

ACKNOWLEDGEMENTS

I am greatly indebted to Dr. M. Wijnans for his indispensable, statistical help, to Prof. Dr. E. M. Cohen, Prof. Dr. P. Sevenster, and Prof. Dr. J. J. A. van Iersel for their stimulating interest and valuable criticism, and to Miss T. P. Frerichs and Miss M. Remmelts for their excellent technical assistance and the patience they had with the animals and myself.

REFERENCES

BUREŠ, J., BUREŠOVÁ, O., BOHDANECKÝ, Z. AND WEISS, T. (1964) The effect of physostigmine and atropine on the mechanism of learning. In *Animal Behaviour and Drug Action*, Churchill, London, pp. 134–143.

DE JONGE, H. (1964) Inleiding tot de medische statistiek, dl. II. Verhandeling van het Nederlands Instituut voor Praeventieve Geneeskunde.

DEMBER, W. N. AND FOWLER, H. (1958) Spontaneous alternation behavior. *Psychol. Bull.*, **55**, 412–428.
DOUGLAS, R. J. (1966) Cues for spontaneous alternation. *J. comp. physiol. Psychol.*, **62**, 171–183.
DOUGLAS, R. J. AND ISAACSON, R. L. (1965) Homogeneity of single trial response tendencies and spontaneous alternation in the T-maze. *Psychol. Rep.*, **16**, 87–92.
DOUGLAS, R. J. AND ISAACSON, R. L. (1966) Spontaneous alternation and scopolamine. *Psychon. Sci.*, **4**, 283–284.
GRANDJEAN, E. AND BÄTTIG, K. (1962) Die Wirkung verschiedener Psychopharmaka auf die spontane Alternation der Ratte. *Helv. physiol. pharmacol. Acta*, **20**, 373–381.
HARTER, H. L. (1960 a) Critical values for Duncan's new multiple range test. *Biometrics*, **16**, 671–685.
HARTER, H. L. (1960 b) Tables of range and studentized range. *Ann. math. Statist.*, **31**, 1122–1147.
KOPPENAAL, R. J. (1962) On the determination of expected probabilities of alternation. *Psychol. Rep.*, **11**, 666.
MEYERS, B. AND DOMINO, E. F. (1964) The effect of cholinergic blocking drugs on spontaneous alternation in rats. *Arch. int. Pharmacodyn.*, **150**, 525–529.
NATRELLA, M. G. (1963) *Experimental Statistics*, National Bureau of Standards, Handbook 91, pp. 13–30, 13–35.
OLIVERIO, A. (1968) Effects of scopolamine on avoidance conditioning and habituation of mice. *Psychopharmacologia (Berl.)*, **12**, 214–226.
OVERTON, D. A. (1966). State dependent learning produced by depressant and atropine-like drugs. *Psychopharmacologia (Berl.)*, **10**, 6–31.
OWEN, D. B. (1962) *Handbook of Statistical Tables*, Addison-Wesley, Reading, Mass.
PARKES, M. W. (1965) An examination of central actions characteristic of scopolamine: comparison of central and peripheral activity in scopolamine, atropine and some synthetic basic esters. *Psychopharmacologia (Berl.)*, **7**, 1–19.
SIEGEL, S. (1956) *Non-parametric Statistics*. McGraw-Hill, New York.
SINHA, S. N., FRANCKS, C. M. AND BROADHURST, P. L. (1958) The effect of a stimulant and a depressant drug on a measure of reactive inhibition. *J. exp. Psychol.*, **56**, 349–354.
SQUIRE, L. R. (1969) Effects of pretrial and posttrial administration of cholinergic and anticholinergic drugs on spontaneous alternation. *J. comp. physiol. Psychol.*, **69**, 69–75.
STILL, A. W. (1966) Memory and spontaneous alternation in the rat. *Nature (Lond.)*, **210**, 400–401.
WALKER, E. L. (1956) The duration and course of the reaction decrement and the influence of reward. *J. comp. physiol., Psychol.*, **49**, 167–176.
YATES, F. (1937) The design and analysis of factorial experiments. Technical Communication No. 35, Imperial Bureau of Soil Science, Hampden, England.
ZAHNER, H., BÄTTIG, K. AND GRANDJEAN, E. (1961) Das spontane Links–Rechts Alternieren der Ratte und dessen Beeinflussung durch Psychopharmaka. *Helv. physiol. pharmacol. Acta*, **19**, 38–39.
ZEAMAN, D. AND HOUSE, B. J. (1951) The growth and decay of reactive inhibition as measured by alternation behavior. *J. exp. Psychol.*, **41**, 177–186.

DISCUSSION

KING: Your paper deals with alternation behaviour largely in terms of stimuli involved. However, as you are no doubt aware, other explanations in terms of the responses made, particularly their direction (left *vs* right) were proposed some time ago.

VAN DER POEL: You are quite right in saying that alternation could be alternation between responses or alternation between stimuli, The former possibility can be disregarded since the work of Douglas (1966). The latter possibility has been stressed by Glanzer (1953) and is known as the "stimulus satiation"—or "habituation"—hypothesis. Strictly speaking, this hypothesis implies that a certain set of stimuli, impinging on the animal at the choice point, cannot induce a certain response any more.

Sutherland (1957) has shown that an experienced animal avoids stimuli beyond the choice point which it has recently perceived, but which do not impinge on the animal at the choice point. I may add 3 observations supporting the latter hypothesis. Firstly, during the first trial of each new day most animals perform Intention movements, thereby exposing themselves to both sets of stimuli (left and right). These animals alternate at least as well as animals which do not show Intention movements.

Further, completely naive rats performing their very first couple of trials in a T-maze, alternate at chance level. Pre-exposure to the T-maze without opportunity to alternate, has a marked positive effect on this first day alternation, indicating that latent learning is involved. In other words, in order to be able to alternate rats have to know "what is leading to what" in the T-maze. Finally, some rats show by slight deviations of the direction of running as soon as they leave the startbox, whether they will choose left or right. So, at least for these animals stimuli arising from the startbox and the first part of the leg of the T, are sufficient to induce an alternation response.

In my opinion, the basis for the alternation is that a certain expectation as to which direction to choose is generated as soon as the animal is released from the startbox. On the basis of this expectation the rat begins to move. During the choice time, virtually at the choice point, this expectation is constantly compared with newly entering information and corrections are made if necessary. I do not think that a simple habituation type of memory is involved in alternation.

RICK: In Professor Kerkut's experiments which he described yesterday, the effect of the AChE was to decrease the learning time and, essentially, in your terminology, this would increase the rate of registration. You showed that the cholinolytic which is going to break up the ACh system somewhat decreases the ability of the animal to register, so to some extent one could agree that these two pieces of work are in tune with one another.

VAN DER POEL: May I try to complete the picture for the alternation behaviour of rats? Squire (1969), who did not train his animals and still used intertrial intervals of more than 1 h (with these long intervals low alternation percentages are observed), was able to demonstrate that a dose of 0.4 mg/kg physostigmine caused an increase in the percentage of animals alternating.

REFERENCES

DOUGLAS, R. J. (1966) Cues for spontaneous alternation. *J. comp. physiol. Psychol.*, **62**, 171–183.

GLANZER, M. (1953) The role of stimulus satiation in spontaneous alternation. *J. exp. Psychol.*, **45**, 387–393.

SQUIRE, L. R. (1969) Effects of pretrial and posttrial administration of cholinergic and anticholinergic drugs on spontaneous alternation. *J. comp. physiol. Psychol.*, **69**, 69–75.

SUTHERLAND, N. S. (1957) Spontaneous alternation and stimulus avoidance. *J. comp. physiol. Psychol.*, **50**, 358–362.

Central Cholinergic Mechanisms in the Thermoregulation of the Rat

E. MEETER

Medical Biological Laboratory TNO, Rijswijk Z.H. (The Netherlands)

If rats, during a brief hexobarbitone anaesthesia, are intravenously injected with barely sublethal doses of organo-phosphorus cholinesterase inhibitors which are able to pass the blood–brain barrier, the animals produce a hypothermia of 4–6° in 2–3 h, followed by spontaneous recovery in 12–20 h. This phenomenon has also been demonstrated in mice but not in guinea pig or rabbit. A few clinical reports of human cases of organo-phosphate poisoning mention a severe drop in the body temperature of the victims. The anticholinesterase hypothermia in the rat can partly be prevented by systemic atropine, but not by atropine methyl nitrate (for details and references, see Meeter, 1971a).

It could be shown that the hypothermia is caused by a combination of increased heat loss and reduced heat production. As to heat loss, the anticholinesterases appear to shift the set-point for heat release (SPHR) of the hypothalamic thermostat to a lower level. This makes the rat use its facilities for releasing heat (vasodilatation in tail and feet) until the body temperature has again become lower than the new SPHR. The lowered set-point can be restored completely to the normal level by intracerebroventricular (i.c.v.) application of atropine (Meeter, 1971b).

Intraventricular injection of carbachol produces a hypothermia which is similar to that following anticholinesterase administration, but is markedly shorter in duration and can be repeated in one rat every 1 or 2 h. It appeared that under suitable experimental conditions the animals produce no vasodilatation following an i.c.v. carbachol injection. This occurs when, due to prolonged inactivity in a cool environment, the body temperature has gone down sufficiently before the carbachol is given. Under such conditions the effect of carbachol on heat production can be studied separately. By recording oxygen consumption as well as body temperature in the resting rat, it could be shown that immediately following the i.c.v. carbachol administration, the O_2 consumption falls steeply to about half its previous level (Meeter, 1971c). The rate of fall of the body temperature and the amount of reduction of O_2 consumption are largely independent of the dose of carbachol (0.3 to 5.0 μg), whereas the depth of the hypothermia and the time it takes for the O_2 consumption to regain its normal level, are determined by the dose.

No hormonal factors appeared to be involved in the effect of i.c.v. carbachol on the heat production, as hypophysectomy and adrenalectomy had essentially no effect on the vasodilatation-free hypothermia. The phenomenon is independent of the activity of the striated muscles and the respiration since deep curarization of the rats had no effect. So far, all procedures directed towards reducing or eliminating the function of the liver cells have reduced or eliminated this part of the carbachol hypothermia.

The results of the experiments suggest that in the hypothalamic thermostat of the rat, cholinergic synapses are involved at a high organizational level in all actions aimed at lowering the body temperature.

REFERENCES

MEETER, E. (1971a) Some new effects of anticholinesterases in the whole animal, with special emphasis on the hypothermia inducing action in the rat. In *Mechanisms of Toxicity*. W. N. ALDRIDGE (Ed.), Macmillan, London, pp. 29–38.

MEETER, E. (1971b) The effect of atropine on the hypothermia and the shift in set-point for heat release evoked by a cholinesterase inhibitor in the rat. *Proc. Koninklijke Nederlandse Akademie van Wetenschappen, Series C*, **74**, 105–112.

MEETER, E. (1971c) The mechanism of action of intraventricular carbachol on the body temperature of the rat. *Arch. int. Pharmacodyn.*, **194**, 318–321.

DISCUSSION

CROSSLAND: If you wish to exclude the liver in a conscious rat, it is possible, at one operation, partially to occlude the hepatic vein. At the same time, loose ligatures are placed around the hepatic and portal veins. The ends of the ligatures are exteriorized by bringing them through the animal's flank. After a few weeks, a collateral circulation is established that results in a large amount of the blood by-passing the liver. The loose ligatures can then be tightened by pulling on the strings. The liver is immediately excluded, there is no circulatory shock and the animal survives for several hours.

MEETER: Thank you for that very useful suggestion. This technique will make it possible to study the effect of carbachol with and without liver function in one rat.

SZERB: There should be a peripheral component in the reduction of heat production by the liver, probably mediated by the parasympathetic system. This should be blocked by the methyl atropine.

MEETER: I agree that there must be a peripheral component. I mentioned that atropine sulphate is unable to prevent the anticholinesterase hypothermia completely; in fact it reduces the lowering of the body temperature only by about 50%. In a recent experiment we administered soman intraventricularly in order to obtain a purely central hypothermia. It appeared that this hypothermia could be prevented completely by intraventricular as well as by intraperitoneal atropine sulphate. How methyl atropine interacts with the peripheral component is still unknown; I intend to study that in the future.

BARSTAD: As the tail of the rat is an important heat radiator, what will be the effect on the anticholinesterase hypothermia if you remove the tail?

MEETER: Removal of the tail reduces the hypothermia by about 25%. The tail is indeed one of the main radiators for release of excess heat by vasodilatation, but the skin of the feet is also involved.

BARSTAD: I wonder if other parts of the surface of the animal would take over when you cut the tail off.

MEETER: I have no knowledge of experiments of this sort.

POLAK: I was surprised that the fall of the set-point for heat release induced by soman, is restored during the next few hours. I wondered how this is possible if we consider that soman is an irreversible cholinesterase inhibitor.

MEETER: There is evidence of various kinds that if you treat cholinergic synapses, for example motor end-plates, with a really irreversible cholinesterase inhibitor the function recovers much more rapidly that the enzyme activity. As to the effect on thermoregulation, we may safely assume that only part of the cholinesterase is inhibited by the soman, so that apart from the rate of recovery of the function of individual synapses we may in this case also have to do with a take over by synapses that have suffered less than others from the inhibitor.

RICK: There is, I believe, a species-specific effect in hypothermia and hyperthermia. This has been shown by Feldberg and Myers (1964) and others in their work with catecholamines. To what extent is this species-specific effect on thermoregulation present with anticholinesterases?

MEETER: As stated in the abstract, there is evidence that rat and mouse react in a similar way. In rabbit and guinea pig we could find no effect. On the basis of evidence in the literature, it seems unlikely that the cat and dog react like the rat. Man, on the other hand, seems to react to a severe anticholinesterase intoxication with a hypothermia.

CROSSLAND: I would add that histamine causes hypothermia in rats when it is given by intraventricular injection. The extent of the hypothermia is as large as that produced by carbachol.

MEETER: I did not study histamine, but I might add some details about noradrenaline. A small series of experiments was done in which 5 μg of adrenaline was given intraventricularly and compared with the effect of 2 μg of carbachol. Both substances produced a fall of set-point for heat release of 2–2.5 °C. However, the effect of noradrenaline was followed by a very characteristic overshoot, that is, the set-point went through an abnormally high level before establishing at the normal level again.

REFERENCES

FELDBERG, W. AND MYERS, R. D. (1964) Effects on temperature of amines injected into the central ventricles. A new concept of temperature regulation. *J. Physiol. (Lond.)*, **173**, 226–237.

The Effects of Anticholinergic Drugs, Chlorpromazine and LSD-25 on Evoked Potentials, EEG and Behaviour

D. M. GREEN AND F. A. B. ALDOUS

Chemical Defence Establishment, Porton Down, Salisbury, Wiltshire (Great Britain)

Anticholinergic drugs administered to animals produce patterns in the electroencephalogram (EEG) of high amplitude, low frequency waves normally associated with sleep, but these drugs do not produce grossly overt changes in behaviour. This phenomenon was described by Wikler (1952) as a "pharmacological dissociation between EEG and behaviour". However this "dissociation" reported on animal studies has not been substantiated in clinical studies (Longo, 1966). Objective studies in animals on correlations between electrical activity of the brain and changes in behaviour after administration of anticholinergic drugs therefore demand the use of subtle behavioural experimental procedures carried out simultaneously with recording of the EEG and use of more specific techniques such as the recording of evoked potentials.

In this paper an attempt has been made to study the effect of 3 anticholinergic drugs, namely, atropine, hyoscine and N-methyl-3-piperidyl benzilate, on the EEG and on auditory evoked potentials in cats subjected to different environmental conditions and trained to a conditioned avoidance response. The effects of these drugs have been compared with effects produced by LSD-25 and chlorpromazine.

METHODS

The experiments were carried out on 7 cats. Electrodes were implanted aseptically under nembutal anaesthesia using the methods described by Bradley and Elkes (1953). Stainless steel screw electrodes were sited epidurally over the primary auditory area (AC) of the ectosylvian gyrus and over association areas in the lateral and suprasylvian gyri. An insulated stainless steel monopolar electrode, bared at the tip over a length of 1 mm, was implanted into the lateral lemniscus (LL) using the stereotactic coordinates P4, L3.5 H-8 according to the stereotactic atlas of Snider and Niemer (1961). (After the conclusion of a series of experiments, confirmation of the siting of the electrode in the LL was confirmed by killing the animal and perfusing the brain with formalin. Serial sections stained with luxol blue and cresyl violet were examined microscopically for electrode track and tip location.) Platinum wire electrodes were inserted into the neck muscles in order to record electromyograms. After completion

References pp. 156–157

of the operative procedures the animals were given intramuscular benzathine penicillin and experiments were carried out not earlier than 3 weeks after the operation.

The cats were placed in a sound attenuated chamber approximately 1 m^3 in dimensions where they were observed by closed circuit television. Auditory stimuli were transmitted by a loudspeaker mounted in the ceiling of the chamber and the EEG was recorded on an Elema Mingograph. Evoked potentials (EPs) were led off from the final amplifier stage of the Mingograph into a Biomac 1000 computer of average transients, using a sweep time of 320 msec. Four channels of the Biomac 1000 were used simultaneously, the distance between analysed points being 1.28 msec. EPs in AC were obtained by recording from the active electrode sited in the primary auditory area against an indifferent electrode sited just outside this area. EPs in LL were recorded against the average potential of all other electrode positions by use of the "indifferent" selector on the Elema Mingograph.

Method 1

Auditory stimuli of 30 msec tonal pips (600 Hz, 80 dB intensity) were transmitted to the animal under different environmental conditions and the experiment was divided into 5 recording sessions. In the first session recordings were made under soundproof conditions and the next session was carried out under distractive conditions which was achieved by replacing the solid door of the chamber with a thin perspex panel so that the animal could be distracted by extraneous laboratory noise and human presence. A control injection of normal saline was then given before the third session which was again carried out under soundproof conditions. The drug was then injected, and 30 min later the fourth, followed by the fifth session took place under soundproof and distractive conditions respectively.

In each session the tonal pips were presented in groups of 4, each pip being separated by an interval of 20 sec. The average response obtained from these 4 stimuli was plotted out from the Biomac 1000 and this procedure was repeated 10 times in each session so that the mean amplitude of 10 evoked potential waveforms (each representing the average of 4 EPs) could be calculated.

Method 2

Two cats were trained in the soundproof chamber to cross from one side of a 9 in. barrier to the other on presentation of a pure tone of 600 Hz of approximately 80 dB intensity for 10 sec. Failure to accomplish the crossing within 4 sec was punished by an electric shock (approximately 40 V, 50 Hz) delivered to the animal from a metal grid floor. Training was carried out on consecutive days until a 100% correct response level of the conditioned avoidance response (CAR) was established. The rate at which the auditory stimulus was presented varied from 1–3 per 5 min.

The experiment was divided into trials each consisting of 10 presentations of the conditioned stimulus (CS). During these trials the EEG was recorded continuously and the average of 10 "on" responses (obtained from 3 electrodes in AC and one in LL) evoked by presentation of the CS was obtained from the Biomac 1000.

Four control trials were first carried out and a further series of trials carried out 30 min, 24 h, and in some cases 48 h after drug administration. During these trials the number of correct responses and failures was noted. After administration of the anticholinergic drugs electrode jelly was applied to the feet to ensure electrical contact with the shocking grid.

The drugs were dissolved in sterile physiological saline and administered by the intramuscular route. The drugs and doses used in this study were atropine sulphate (2.4 mg/kg), hyoscine hydrobromide (0.24 mg/kg), N-methyl-3-piperidyl benzilate hydrochloride (0.5 mg/kg), LSD-25 tartrate (10 μg/kg) and chloropromazine hydrochloride (5 mg/kg).

RESULTS

1. Effects of drugs on EPs in cats subjected to different environmental conditions

The method employed to obtain the results in this series of experiments was basically the same as that carried out previously by Key (1965a) who studied the effect of LSD-25 on the amplitude of individually evoked potentials in the cochlear nucleus. In the present study EPs were recorded in the LL and AC and averaged potentials evoked from consecutive tonal pips were analysed in preference to individual responses. This increased the accuracy of amplitude measurements by increasing the signal to noise ratio, but as the fluctuation in amplitude of EPs was also to be studied only a small averaging sample of 4 was chosen.

a. Effects on EPs in LL

Variations in amplitude of EPs in LL under soundproof and distractive conditions in the untreated cat and after LSD-25 administration were similar to results obtained by Key (1965a). Under soundproof conditions, when the untreated animal was relaxed, the fluctuations in amplitude were small, but under distractive conditions the mean amplitude decreased and the fluctuation in amplitude (shown by the standard deviations in Fig. 3) increased. LSD-25 produced desynchronization in the EEG, general alerting, increase in exploratory behaviour, and under soundproof conditions the degree of fluctuation in amplitude of EPs in LL was similar to that obtained in the untreated animal subjected to distractive conditions. When the cats were subjected to distractive conditions after LSD-25 there was a further increase in fluctuation of amplitude and the mean amplitude was lower than that in the untreated cat under similar conditions.

The 3 anticholinergic drugs produced effects which differed from those of LSD-25. EPs recorded before and after administration of N-methyl-3-piperidyl benzilate are shown in Fig. 1, and a comparison of the results obtained with this drug and LSD-25 is shown in Fig. 2. The anticholinergic drugs synchronized the EEG but no gross changes in behaviour were observed. In most cases the mean amplitude of EPs in LL determined both under soundproof and distractive conditions was smaller than the control means determined in the untreated animal under soundproof conditions, but

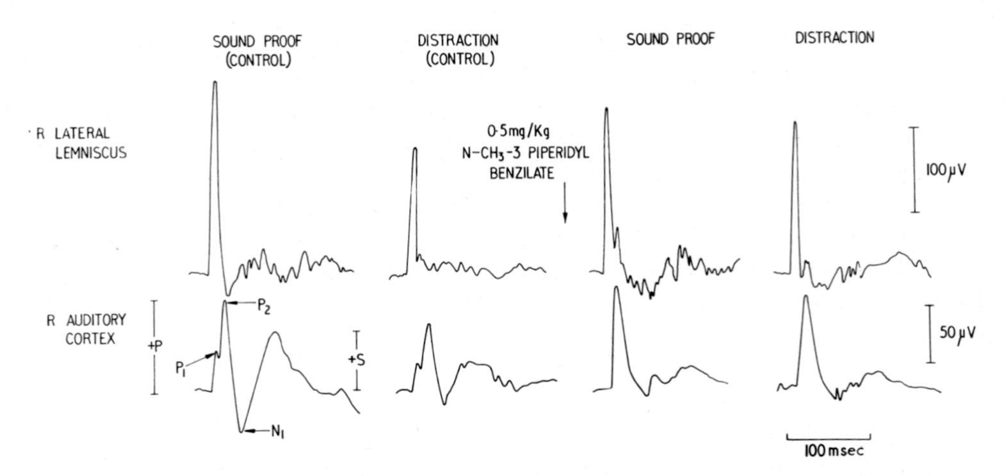

Fig. 1. Records of potentials in the right lateral lemniscus and right auditory cortex evoked by presentation of 30 ms tonal pips (600 Hz) under sound proof and distractive conditions before and after 0.5 mg/kg N-methyl-3-piperidyl benzilate.

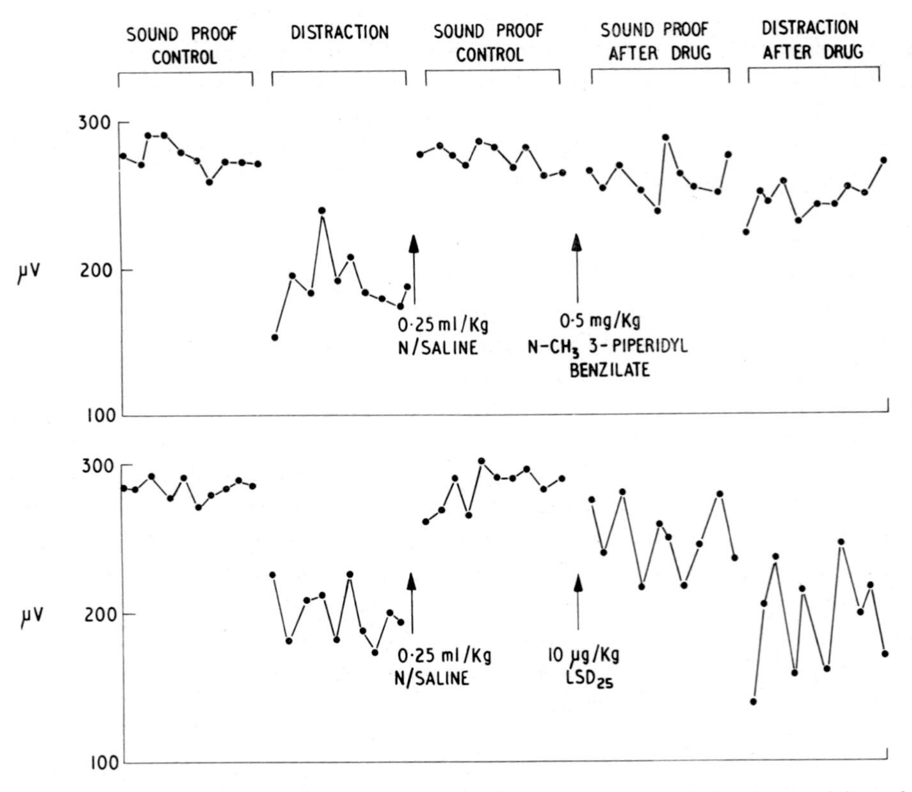

Fig. 2. Effect of N-methyl-3-piperidyl benzilate and LSD-25 on the amplitude of potentials evoked in the lateral lemniscus of the cat in relation to the environmental conditions. Each point on the graphs represents the mean amplitude of 4 individual potentials evoked every 20 s by 30 ms tonal pip stimulation (600 Hz).

greater than the control means determined under distractive conditions. The fluctuation in amplitude never exceeded the level of fluctuation shown in untreated animals subjected to distractive conditions. The results of all the experiments carried out on EPs in LL are summarised in histogram form in Fig. 3, and this figure also shows the effect produced by chlorpromazine. This drug produced its tranquillizing effect on behaviour and synchronized the EEG. The mean amplitude of EPs under both soundproof and distractive conditions was very similar to the mean amplitude determined under soundproof conditions in the untreated animal, but the variation about the means was smaller than the variations about the control means.

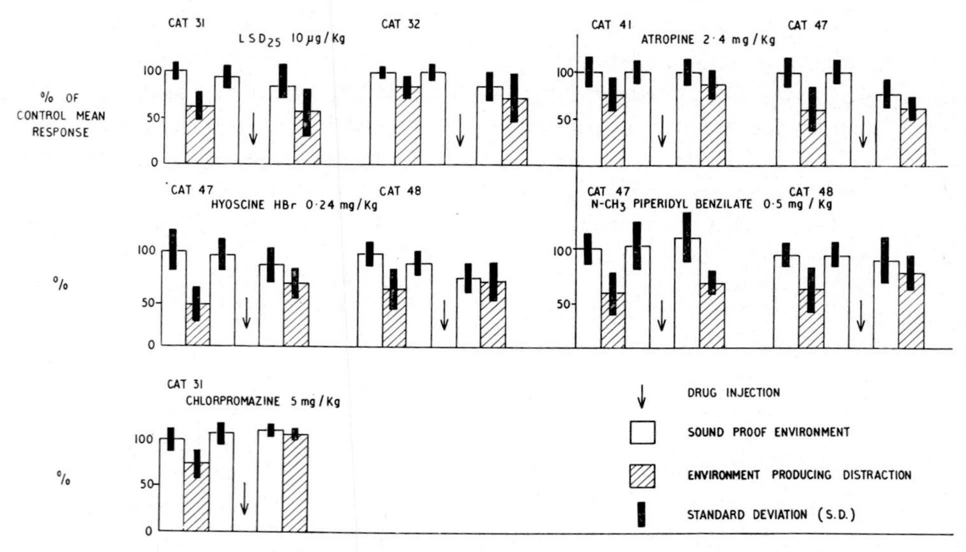

Fig. 3. Histograms showing effects of drugs on the amplitude of potentials evoked in the lateral lemniscus of cats in relation to the environmental conditions.

b. Effects on EPs in AC

The typical surface response evoked in the AC (Fig. 1) consists of two short latency primary potentials (P1 and P2) which are followed by a negative potential (N1) (Teas and Kiang, 1964) and then a longer latency secondary positive potential (+S). The anticholinergic drugs produced a fusion of P1 and P2 potentials and amplitude analysis of these individual waves was therefore impracticable. In this study the maximum height of the primary complex (+P) was analysed together with analysis of the amplitude and peak latency of +S potential.

Changes in the mean amplitude of +P produced by distractive conditions in untreated cats and cats injected with LSD-25 followed a similar pattern to changes that occurred in mean amplitude of EPs in LL. However unlike EPs in LL, fluctuations in +P amplitude under distractive conditions and after LSD-25 were not appreciably larger than fluctuations recorded under control soundproof conditions. The amplitude of the +S potential was extremely variable and no consistent changes could be

References pp. 156–157

detected in the untreated animal subjected to different environmental conditions or in the animal given LSD-25.

Under soundproof and distractive conditions an increase in the mean amplitude of +P and a decrease in +S amplitude was produced by the anticholinergic drugs. The most marked effect that these drugs produced was an increase in the peak latency of the +S potential. An attempt was made to gain information on the significance of these changes by using the second experimental procedure.

2. *Effects of drugs on EPs recorded during a CAR*

Atropine, hyoscine and chlorpromazine administered at the previously stated dose levels produced disruption of the CAR and the results of the experiments are shown graphically in Fig. 4. Chlorpromazine produced a smooth rapid extinction of the CAR, whereas the decrease in performance produced by the anticholinergic drugs was irregular. LSD-25 did not produce a decrease in performance but on the other hand an increase in performance occurred in that this drug produced a significant decrease ($P < .001$) in avoidance latency. The mean of 40 avoidance latency values

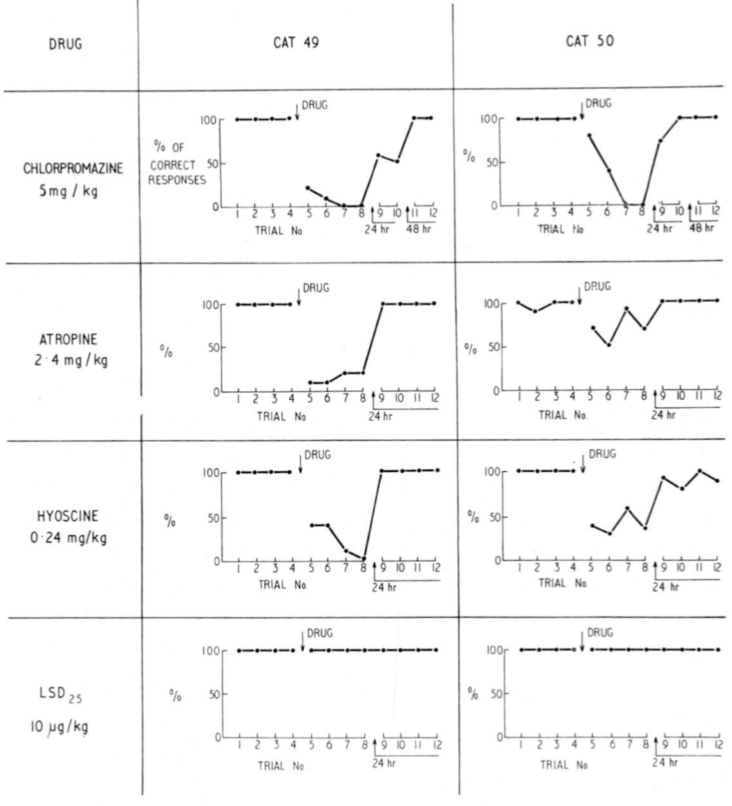

Fig. 4. Effects of drugs on two cats which had been fully trained to a conditioned avoidance response.

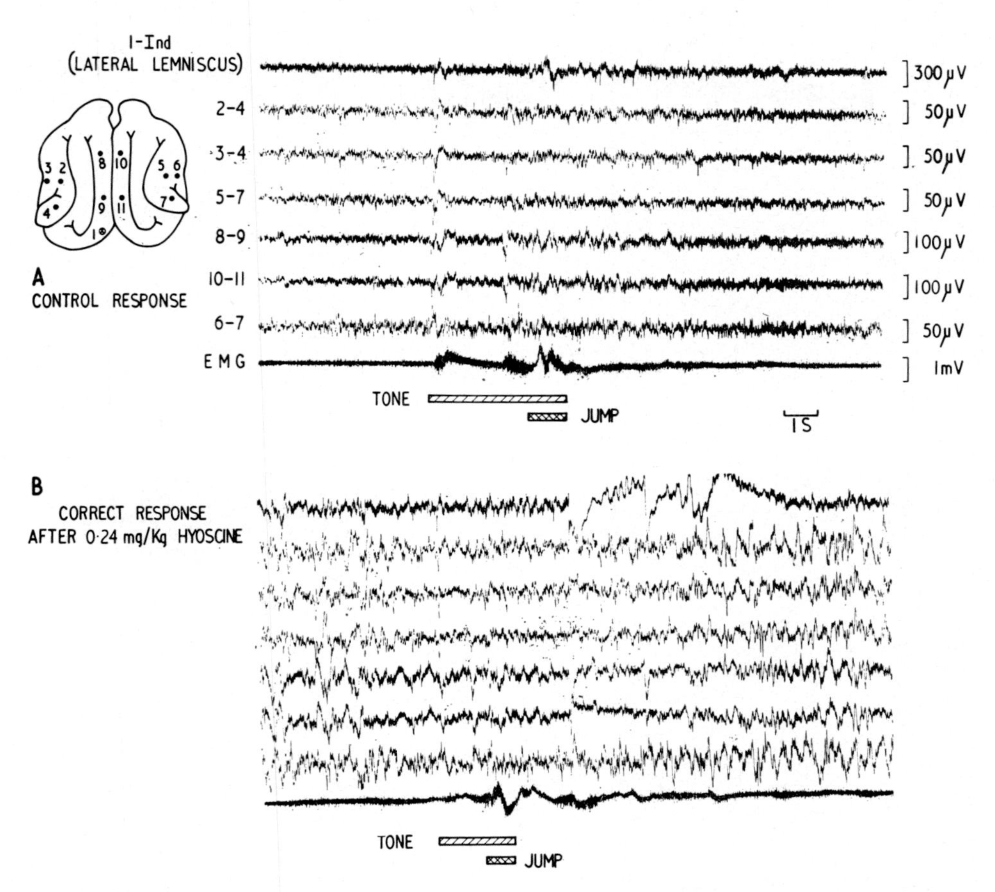

Fig. 5. Examples of EEG records obtained during a conditioned avoidance response in A, an untreated cat and in B, the same cat dosed with hyoscine.

in one untreated cat was 2.78 s and for the other cat 3.46 s. After LSD-25 the mean avoidance latencies were decreased to 1.52 s and 2.29 s respectively.

The EEG of untreated cats during the test procedure was always desynchronized corresponding with the degree of alertness or attentiveness of the animal, and a record obtained during a control CAR is shown in Fig. 5A. The anticholinergic drugs and chlorpromazine synchronized the EEG, and Fig. 5B shows a record obtained during a correct CAR after administration of hyoscine. EPs in the untreated animal were therefore recorded on a desynchronized EEG background whereas EPs recorded after administration of the anticholinergic drugs and chlorpromazine were recorded on a synchronized "sleep-like" EEG background. It is now well established that the waveform of EPs in the normal animal varies with the type of EEG background that they are superimposed upon, for example in the awake–sleep cycle (Herz *et al.*, 1966); bearing this in mind changes in EPs after administration of drugs that alter the EEG must be interpreted with a certain degree of caution (see DISCUSSION).

Changes in EPs recorded from LL and AC during the CAR were compared with

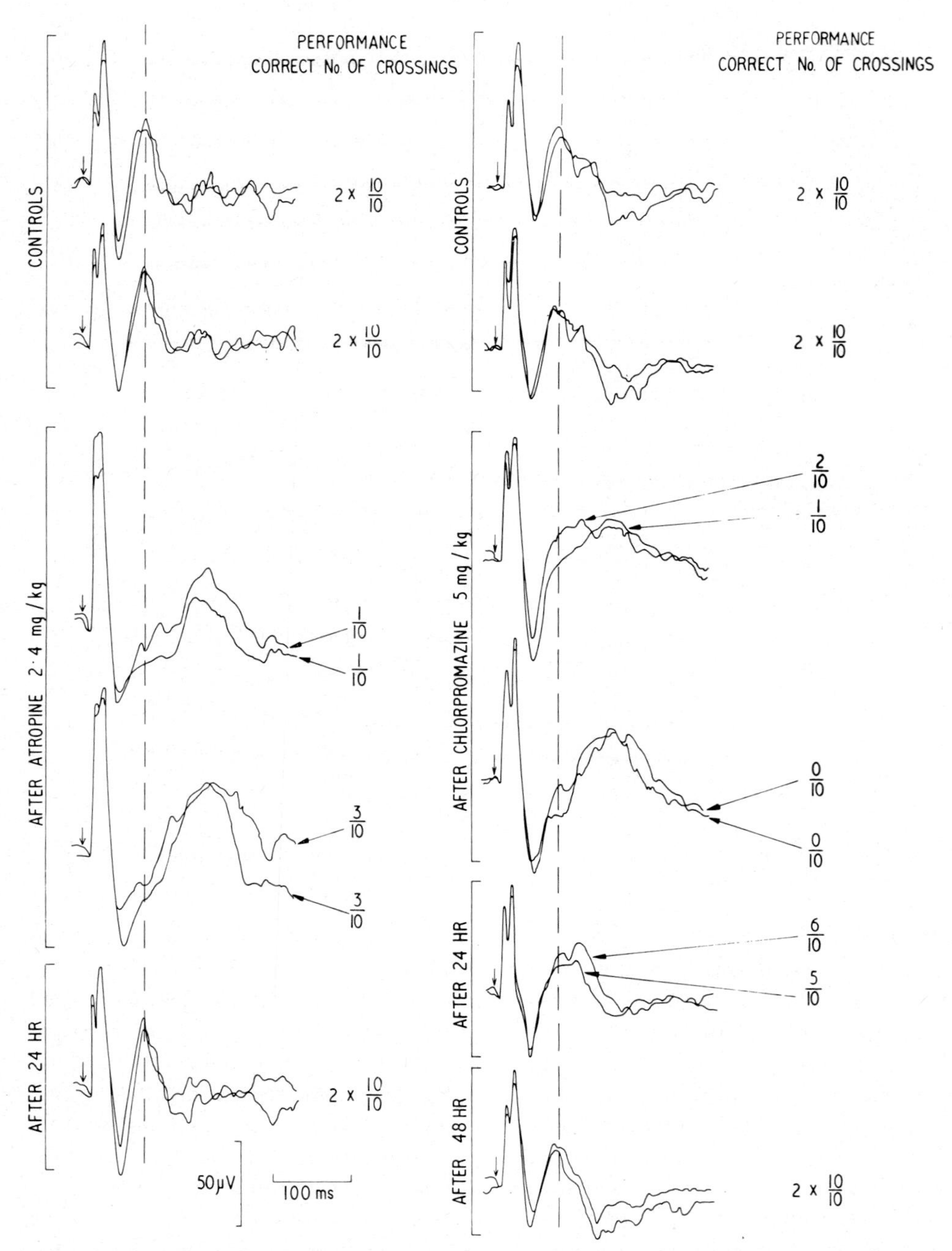

Fig. 6. Comparison of the effects of atropine and chlorpromazine on potentials evoked in the auditory cortex recorded during a conditioned avoidance response test procedure. Each potential waveform represents the average of 10 "ON" responses evoked in one trial by 10 presentations of the conditioned stimulus (600 Hz tone); each potential waveform is related to the number of correct responses obtained in one trial. (↓) presentation of tone; (- - -) peak latency of secondary potential in control animals.

changes in CAR performance. The average EP waveform from 10 CARs in one trial was obtained. Four control trials were carried out giving 4 average waveforms for each electrode position and from these waveforms mean values were obtained for the amplitude of EPs in LL, amplitude of the +P and +S of EPs in AC together with mean values for peak latency of +S. Exactly the same procedure was carried out to determine the mean values after administration of the drug and these means were compared with their respective control means by use of Students *t*-test.

Comparison of the effects of atropine and chlorpromazine on EPs in AC recorded during the CAR trials are shown in Fig. 6, and Fig. 7 summarizes the results of all the experiments. The anticholinergic drugs and chlorpromazine increased the amplitude of the +P potential and the amplitude of +S was decreased. There was however a difference between the effects of chlorpromazine and the anticholinergic drugs in that no fusion between the P1 and P2 potentials was produced by chlorpromazine. Also, there was a tendency for the amplitude of N1 potential to decrease with the anticholinergic drugs but chlorpromazine generally produced the opposite effect. The effects of the drugs on the N1 potential in this study were not consistent and full analysis of changes in amplitude of N1 was not carried out especially as there appears to be some controversy on the electrophysiological mechanisms and pathways involved in the production of this potential (Ochs, 1968). The most significant change produced by the anticholinergic drugs and chlorpromazine was to increase the peak latency of +S. In untreated cats the mean peak latency of +S determined

CAT No.	DRUG	EVOKED POTENTIALS										BEHAVIOUR % No. CORRECT RESPONSES
		LATERAL LEMNISCUS (AMPLITUDE)	AUDITORY CORTEX									
			PRIMARY (+P) AMPLITUDE			SECONDARY (+S)						
						AMPLITUDE			TIME TO PEAK RESPONSE			
			ELECTRODE No.			ELECTRODE No.			ELECTRODE No.			
			1	2	3	1	2	3	1	2	3	
49	ATROPINE 2·4 mg/Kg	↑	↑x	↑+	↑+	↓	↓	↓	↑*	↑*	↑*	20 %
50	"	↑	↑	↑	↑	↓*	↓	↓+	↑*	↑*	↑*	70 %
49	HYOSCINE 0·24 mg/Kg	↑	↑	↑+	↑	↓	↓+	↑	↑*	↑x	↑*	22 %
50	"	↑	↑	↑	↑x	↓+	↓+	↓	↑*	↑*	↑*	42 %
49	CHLORPROMAZINE 5mg/Kg	↓	↑	↑+	↑	↓	↓+	↓	↑*	↑x	↑x	8 %
50	"	↓	↑	↑	↑	↓	↓+	↓	↑*	↑x	↑*	25 %
49	LSD 25 10 µg/Kg	↑x	↓	↓	↑	↓	↑	↑	↑	↑	↓	100 %
50	"	↑+		↓	↓x		↓	↑		↑	↓	100 %

↑ INCREASE

↓ DECREASE

* $P < \cdot 001$

X $P < \cdot 01$

\+ $P < \cdot 05$

Fig. 7. Summary of results of the effects of drugs on behaviour and evoked potentials recorded from cats during a conditioned avoidance response test procedure.

from a total of 6 different electrodes sited in the AC was 84 ms and after administration of atropine, hyoscine and chlorpromazine the mean peak latency was increased to 162 ms, 161 ms and 131 ms respectively. It was interesting to note that in one cat when the CAR recovered to the 50% correct response level 24 h after administration of chlorpromazine the peak latency was 97 msec which was still significantly different ($P < 0.01$) from the control mean of 84 ms. This effect is shown in Fig. 6.

The only common change produced by the anticholinergic drugs and chlorpromazine on EPs in AC was therefore a reduction in the amplitude and increase in peak latency of +S; these changes were accompanied by a decrease in CAR performance. Both atropine and hyoscine increased the amplitude of EPs in LL during the CAR but this increase was not significant at the $P < 0.05$ level. Chlorpromazine on the other hand produced a decrease in amplitude but again this was not significant.

There were no consistent changes produced in EPs in AC during the CAR after administration of LSD-25 and the only changes in EPs that could possibly be associated with the decrease in avoidance latency was a significant increase ($P < 0.05$) in the amplitude of EPs in LL.

DISCUSSION

The results described in this paper are based on one dose level for each drug studied and it seems necessary first to discuss the choice of dose used. In order to determine that changes in behaviour may be associated with slow wave activity in the EEG produced by administration of anticholinergic drugs a dose level had to be chosen that could produce slow waves in the EEG but would not produce lack of motor coordination. Doses used in this study were twice those required to elevate EEG arousal threshold by 100% (Brimblecombe and Green, 1968). The doses of LSD-25 and chlorpromazine were similar to those used by Key and Bradley (1960) in a comparison of the effects of these drugs on conditioning and habituation to arousal stimuli in cats.

Studies in man have shown that central effects produced by atropine and hyoscine are similar to those produced by the piperidyl benzilates (Longo, 1966). Pfeiffer (1959) studied the central effects in man of some central anticholinergic drugs including N-methyl-3-piperidyl benzilate (JB 366) and found that these drugs produced an effect which the subjects likened to LSD-25. On the other hand Cerletti (1963) emphasized differences between the psychotomimetic effects produced by the anticholinergic piperidyl benzilates and those produced by LSD-25. For example, the piperidyl benzilates produced a misinterpretation and isolation from the environment with a tendency to introversion whereas LSD-25 produced an extrovertive attitude facilitating psychotherapeutic procedures.

Studies in animals with auditory evoked potentials have shown that fluctuations in amplitude may be related to the significance attached to the stimulus, for if a stimulus is made more meaningful the potential tends to increase in size (Galambos, 1960). Environmental distraction is shown by a reduction in mean amplitude of EPs (Hernández–Peón *et al.*, 1956) but fluctuation in amplitude is generally increased (Key, 1965a). It has been suggested from behavioural studies (Key and Bradley, 1960; Key, 1961) that LSD-25 increases the significance of stimuli, previously of less significance, which in turn increases the degree of distractibility; this increase in distractibility is thus associated with an increased fluctuation in amplitude of EPs. These effects have been confirmed in the present paper by studying changes in the amplitude of EPs in the LL. The finding that after administration of anticholinergic drugs the fluctuation in amplitude of EPs in LL recorded under distractive conditions was smaller than the fluctuation recorded under similar conditions in the untreated animal suggest that these drugs decrease the distractibility of the animal. This decrease in distractibility by the anticholinergic drugs and increase in distractibility produced by LSD-25 may be associated respectively with the "introvertive" and "extrovertive" states produced in man.

Further confirmation that the degree of distractibility is dependent on the fluctuation in amplitude of EPs in LL is provided by the results obtained with the tranquillizing drug chlorpromazine when only very small fluctuations in amplitude occurred with this drug under both soundproof and distractive conditions.

As consecutive auditory stimuli were used in Method 1, a certain degree of habituation to the stimuli may have occurred. Amplitudes of EPs in the cochlear nucleus are not reduced during habituation (Marsh and Worden, 1964) and thus it is unlikely that amplitudes of EPs in the LL are affected by this phenomenon. However a marked reduction in amplitude occurs during habituation in primary responses recorded in AC (Marsh and Worden, 1964) and this may partly explain the smaller degree of amplitude fluctuation seen in primary AC responses than in responses evoked in the LL. Another important factor to be considered in the analysis of EPs in AC is the modulation of EP waveforms by the EEG background pattern. During drowsiness and during high voltage slow wave sleep (HVS sleep) the amplitude of primary responses evoked in AC are larger than in the alert animal when the EEG is desynchronized (Herz *et al.*, 1966). In order to determine, therefore, that observed changes in EPs are produced by a specific action of the drug and not due to a modulation effect by EEG background activity EPs recorded after drug administration should be compared, if possible, with EPs recorded from untreated animals when the EEG background is similar to that produced by the drug. The anticholinergic drugs produced HVS sleep-like activity in the EEG, although the animals remained behaviourally alert, and the fusion that occurred between the P1 and P2 potentials of the primary evoked response in AC confirmed the findings of Herz *et al.* (1967). The fusion between these potentials could be attributed to a specific action of the anticholinergic drugs as no fusion occurs during normal HVS sleep. The P1 wave reflects the arrival of the impulse through the primary sensory afferent pathway and excitation of neurones in the deeper layer of the cortex (Teas and Kiang, 1964) whereas there is controversy on the pathways involved in the production of the P2 wave. The most feasible explanation appears to be that put forward by Anochin (1961) that P2 is produced by several corticopetal pathways, and thus the fusion of

References pp. 156–157

P1 and P2 into a single wave points to an effect of anticholinergic drugs at the cortical level.

Of particular interest was the finding that the anticholinergic drugs reduced the amplitude and increased the peak latency of the +S potential. Szerb (1965) studying average evoked potentials in the somatosensory cortex of cats anaesthetized with halothane or Dial found that atropine 0.2 mg/kg i.v. and 0.1 mg/ml, applied to the cortex directly, decreased the mean amplitude of the secondary potential. Studies in our laboratory have shown that anticholinergic drugs produced a significant decrease in amplitude and increase in peak latency of +S when compared with EPs in AC recorded during natural HVS sleep. These changes produced in the +S potential are therefore due to a specific action of the anticholinergic drugs and not due to modulation by EEG background activity.

There is general agreement in the literature (Goff, 1970) that in clinical studies the so-called late components of averaged cerebral responses, those occurring later than 80–100 ms, are the most labile under experimental operations designed to alter attention or the perceptual organisation of the subject. Key (1965b) showed that in cats the amplitude of the +S component of EPs in AC, rather than that of the primary component, was closely related to the significance of the stimulus at least in terms of its arousal or attentive value. The results obtained in this paper on EPs evoked by consecutive stimuli under different environmental conditions suffer from the experimenter not knowing the level of significance or meaningfulness that the animal attaches to the stimulus, and for a more detailed analysis of drugs affecting the EPs significance was attached to the stimulus by making it a conditioned stimulus as part of a CAR (Method 2).

The results show that with the anticholinergic drugs and chlorpromazine, reduction in amplitude and increase in peak latency of the +S potential is associated with decrease in performance in the CAR; these findings imply that the changes in +S potential are related to the changes in significance or meaningfulness of the stimulus. The amplitude of the +P potential was generally increased but as these drugs produced HVS sleep-like activity in the EEG this increase in amplitude may have been due to modulation of the EP by EEG background activity.

The +S potential could be the result of the arrival of impulses in deeper layers of the cortex mediated through the brain stem reticular formation and non-specific thalamic structures. Although the anticholinergic drugs and chlorpromazine produced similar effects on the +S potential the site of action producing these effects is probably different. The effect of topically applied atropine (Szerb, 1965) in reducing the amplitude of the long latency secondary potential in the somatosensory cortex suggests that the anticholinergics affect secondary potentials at the cortical level, whereas chlorpromazine may affect these potentials by inhibiting collaterals which connect sensory pathways with the non-specific system (Bradley, 1963; Key, 1965c). It has been mentioned (Bradley and Hance, 1957) that chlorpromazine itself possesses weak anticholinergic activity and reduction in amplitude of +S potential may therefore be due to this type of activity; this is not likely as chlorpromazine did not produce a fusion of P1 and P2 potentials and the amplitude of N1 tended to increase rather than decrease.

It must be borne in mind that physiological studies have shown (Goff, 1970) that it is possible to record late potentials where only the specific system is present after destruction of the non-specific pathways. From these studies it is suggested that late evoked potentials may not reflect activity mediated by an independent extralemniscal projection system but appears to be dependent on the integrity of the lemniscal system at the thalamocortical and cortical level.

The effect of LSD-25 in increasing the mean amplitude of EPs in LL during the CAR may be related to the effect of this drug in increasing the significance of the conditioned stimulus which in turn increases the performance as shown by a decrease in avoidance latency. The site of action of LSD-25 is still undefined but it has been suggested that this drug specifically interferes with processes regulating the flow and integration of sensory information within the brain; it is possible that LSD-25 primarily affects information relayed from first or second order neurones in the caudal brain stem areas (Key, 1965c).

While there is almost complete agreement that anticholinergic drugs disrupt insufficiently consolidated responses there are conflicting reports of these drugs disrupting the CAR in fully trained animals. Studies have shown that in rats there is no disruption (Herz, 1968) and in some cases an improved performance has been observed (Longo, 1966). However when higher animals are used there are reports that atropine (1 mg/kg) disrupts the CAR in monkeys (Ricci and Zamparo, 1965) and atropine (3 mg/kg) and hyoscine (0.1 mg/kg) were found to disrupt the running behaviour of dogs performing an exercise avoidance test (Mennear *et al.*, 1965). The action of anticholinergic drugs on previously acquired conditioned responses is therefore very much dependent on the species of animal. Funderburk and Case (1947) found that 1.2 mg/kg atropine did not affect the CAR in cats trained to avoid a shock by crossing a barrier. However, at this dose level atropine does not produce its maximum effect in synchronizing the EEG as the EEG arousal threshold is only raised by 100% over the control value, but when the dose is raised to 2.4 mg/kg a maximal effect is observed when the EEG arousal threshold is raised by 200% (Brimblecombe *et al.*, 1971). The effect of 2.4 mg/kg in disrupting the CAR is reported in this paper.

The finding that anticholinergic drugs in the cat disrupt the CAR together with the production of high voltage slow waves in the EEG demonstrate that there is an "association" and not a "dissociation" between EEG activity and behaviour. Also, the finding that the anticholinergic drugs and chlorpromazine produce a similar synchronizing effect on the EEG but produce different effects on primary evoked potentials in the cortex substantiates the conclusions made by Herz *et al.* (1967) and White *et al.* (1965) that mechanisms which induce changes in the background EEG are different from those which affect evoked potentials.

References pp. 156–157

SUMMARY

1. The effects of 3 anticholinergic drugs (atropine, hyoscine and N-methyl-3-piperidyl benzilate) have been studied on the EEG and potentials evoked in the specific auditory pathway (lateral lemniscus and auditory cortex) in cats subjected to different environmental conditions and trained to a conditioned avoidance response.

2. Fluctuation in amplitude of potentials evoked in the lateral lemniscus appeared to be related to the distractability or attentiveness of the animal. After administration of anticholinergic drugs fluctuation in amplitude of potentials recorded under distractive conditions was smaller than in untreated animals subjected to similar environmental conditions; LSD-25 increased the degree of fluctuation.

3. The anticholinergic drugs and chlorpromazine produced synchronization in the EEG, disruption of the conditoned avoidance response and decrease in amplitude and increase in peak latency of the surface positive secondary potential evoked in the auditory cortex. It is suggested that changes in this secondary potential are related to changes in significance or meaningfulness of the stimulus.

4. The finding that anticholinergic drugs in the cat disrupt the conditioned avoidance response together with the production of slow wave high voltage activity in the EEG demonstrate that there is an "association" and not a "dissociation" between EEG activity and behaviour.

REFERENCES

ANOCHIN, P. K. (1961) The multiple ascending influences of the subcortical centres in the cerebral cortex. In *Brain and Behaviour,* M. A. B. BRAZIER (Ed.), Vol. I, Amer. Inst. Biol. Sci., Washington, pp. 138–170.

BRADLEY, P. B. (1963) Phenothiazine derivatives. In *Physiological Pharmacology*, W. S. ROOT AND F. HOFFMAN (Eds.), Academic Press, New York, pp. 407–472.

BRADLEY, P. B. AND ELKES, J. (1953) Technique for recording the electrical activity of the brain in the conscious animal. *Electroenceph. clin. Neurophysiol.*, **5**, 451–456.

BRADLEY, P. B. AND HANCE, A. J. (1957) The effect of chlorpromazine and methopromazine on the electrical activity of the brain in the cat. *Electroenceph. clin. Neurophysiol.*, **9**, 191–215.

BRIMBLECOMBE, R. W. AND GREEN, D. M. (1968) The peripheral and central actions of some anticholinergic substances. *Int. J. Neuropharmacol.*, **7**, 15–27.

BRIMBLECOMBE, R. W., GREEN, D. M., ALDOUS, F. A. B. AND THOMPSON, P. B. J. (1971) Central and peripheral actions of anticholinergic drugs administered with triflupromazine. *Neuropharmacology*, **10**, 93–101.

CERLETTI, A., SCHLAGER, E., SPITZER, F. AND TAESCHLER, M. (1963) Psychodisleptica. *Schweiz. Apoth.-Ztg.*, **101**, 210–240.

FUNDERBURK, W. H. AND CASE, J. J. (1947) Effects of parasympathetic drugs on the conditioned response. *J. Neurophysiol.*, **10**, 179–187.

GALAMBOS, R. (1960) In *Neural Mechanism of the Auditory and Vestibular Systems*, G. L. RASMUSSEN AND W. R. WINDLE (Eds.), Thomas, Springfield, Ill., p. 137.

GOFF, W. R. (1970) Evoked potential correlates of perceptual organisation in man. In *Attention in Neurophysiology*, C. R. EVANS (Ed.), Butterworths, London, pp. 169–193.

HERNÁNDEZ-PEÓN, R., SCHERRER, H. AND JOUVET, M. (1956) Modification of electrical activity in the cochlear nucleus during "attention" in the unanaesthetized cat. *Science*, **123**, 331–332.

HERZ, A. (1968) Some actions of cholinergic and anticholinergic drugs on reactive behaviour. In *Progr. in Brain Res.*, P. B. BRADLEY AND M. FINK (Eds.), Vol. 28, Elsevier, Amsterdam, pp. 73–85.

HERZ, A., FRALING, F., NEIDNER, I. AND FARBER, G. (1967) Pharmacologically induced alterations of cortical and subcortical evoked potentials compared with physiological changes during the awake–sleep cycle in cats. *Electroenceph. clin. Neurophysiol.*, **26**, 164–176.

HERZ, A., NEIDNER, I., FRALING, F. AND SOMMER–SMITH, I. (1966) Corticale und subcorticale Reaksionspotentiale nach sensorischer Reizung der wachen und schlafenden Katze. *Exp. Brain Res.*, **1**, 249–264.

KEY, B. J. (1961) The effects of drugs on discrimination and sensory generalization of auditory stimuli in cats. *Psychopharmacologia (Berl.)*, **2**, 352–363.

KEY, B. J. (1965 a) Effect of lysergic acid diethylamide on potentials evoked in the specific sensory pathway. *Brit. med. Bull.*, **21**, 30–35.

KEY, B. J. (1965 b) Correlation of behavior with changes in amplitude of cortical potentials evoked during habituation by auditory stimuli. *Nature (Lond.)*, **207**, 441–442.

KEY, B. J. (1965 c) The effects of drugs in relation to the afferent collateral system of the brain stem. *Electroenceph. clin. Neurophysiol.*, **18**, 670–679.

KEY, B. J. AND BRADLEY, P. B. (1960) The effects of drugs on conditioning and habituation to arousal stimuli in animals. *Psychopharmacologia (Berl.)*, **1**, 450–462.

LONGO, V. G. (1966) Behavioural and electroencephalographic effects of atropine and related compounds. *Pharmacol. Rev.*, **18**, 965–996.

MARSH, J. T. AND WORDEN, F. G. (1964) Auditory potentials during acoustic habituation: cochlear nucleus, cerebellum and auditory cortex. *Electroenceph. clin. Neurophysiol.*, **17**, 658–692.

MENNEAR, J. H., SAMUEL, G. K., JOFFE, M. H. AND KODAMA, J. K. (1966) Effects of scopolamine and atropine on the performance of an exercise avoidance test by dogs. *Psychopharmacologia (Berl.)*, **9**, 347–350.

OCHS, S. (1968) *Elements of Neurophysiology*, Wiley, New York, pp. 454–458.

PFEIFFER, C. C. (1959) Parasympathetic neurohumors: possible precursors and effect on behaviour. *Int. Rev. Neurobiol.*, **1**, 195–244.

RICCI, G. F. AND ZAMPARRO, L. (1965) Electrocortical correlates of avoidance conditioning in the monkey, their modification with atropine and amphetamine. In *Pharmacology of Conditioning, Learning and Retention*, M. Y. MICHELSON AND V. G. LONGO (Eds.), Czechoslovak Medical Press, Prague, pp. 269–283.

SNIDER, R. S. AND NIEMER, W. T. (1961) *A Stereotaxic Atlas of the Cat Brain.* Univ. of Chicago Press, Chicago.

SZERB, J. C. (1965) Average evoked potentials and cholinergic synapses in the somatosensory cortex of the cat. *Electroenceph. clin. Neurophysiol.*, **18**, 140–146.

TEAS, D. C. AND KIANG, N. C. (1964) Evoked responses from the auditory cortex. *Exp. Neurol.*, **10**, 91–119.

WIKLER, A. (1952) Pharmacologic dissociation on behaviour and EEG sleep patterns in dogs: morphine, N-allylmorphine, and atropine. *Proc. Soc. exp. Biol. (N.Y.)*, **79**, 261–265.

WHITE, R. P., SEWELL, H. H., JR. AND RUDOLPH, H. S. (1965) Drug induced dissociation between evoked reticular potentials and the EEG. *Electroenceph. clin. Neurophysiol.*, **19**, 16–24.

DISCUSSION

BRADLEY: I am sorry that there still seems to be some misunderstanding about "dissociation" with cholinolytic drugs. What Wikler (1952) described in the dog and we reported for the cat and other species is that there is a pattern of activity in the EEG which resembles that of slow wave sleep but that sleep does not occur behaviourally. We never suggested that there is no change in behaviour after atropine but that the behaviour you might expect from the EEG, that is sleep, does not occur. I am pleased to learn that you have found behavioural changes with atropine, as have others, but this does not represent "association" as you described it.

GREEN: I agree that "dissociation between EEG and behaviour" was originally assessed in basic terms of sleep and wakefulness. Nevertheless, a great deal of controversy has arisen over the term "dissociation" with anticholinergic drugs mainly due to problems of terminology, species differences and differing techniques of observation and quantification.

Bearing these points in mind in considering the results in the present paper, the terms "dissociation" and "association" may be regarded as inappropriate and an alternative conclusion may be simply

Reference p. 158

that anticholinergic drugs produce disruption of the conditioned avoidance response accompanied by synchronised activity in the EEG, but that the alertness of the animal is not grossly affected.

KERKUT: I am not clear on the behavioural side about the link between the effects of anticholinergic drugs on learning and on cutting out extraneous noise. If the animals learn more quickly, you would expect them to pay attention to what is going on. I wonder whether an explanation of the failure of your cats to respond when they are conditioned under atropine is because they are not paying attention to the signal.

GREEN: Under our experimental conditions the changes in the secondary potential could be related to the meaningfulness of the stimulus and in this case your supposition may be correct. It would be interesting to compare the effects of anticholinergic drugs on the secondary potential in naive cats, trained up to the 100% correct level, with the effects of these drugs on this potential in fully trained cats.

REFERENCES

WIKLER, A. (1952) Pharmacologic dissociation of behaviour and EEG sleep patterns in dogs: Morphine, N-allylmorphine and atropine. *Proc. Soc. exp. Biol. (N.Y.)*, **79**, 261–265.

The Effect of Atropine on the Metabolism of Acetylcholine in the Cerebral Cortex

J. C. SZERB

Department of Physiology and Biophysics, Dalhousie University, Halifax (Canada)

There is extensive evidence for the role of acetylcholine (ACh) as a synaptic transmitter in the cerebral cortex. The action of ACh on cortical units is predominantly excitatory and antagonized specifically by atropine and other antimuscarinic agents (Krnjević and Phillis, 1963; Crawford and Curtis, 1966). Inhibition of the action of presynaptically released ACh can be expected to alter the metabolism of ACh in the cortex, thereby revealing mechanisms that control the activity of these cholinergic neurones. Indeed, systemically administered atropine increases the release of ACh after cholinesterase inhibition into physiological saline solutions kept in contact with the pial surface of the cortex (Mitchell, 1963; MacIntosh, 1963). Since application of small amounts of atropine to the cortex similarly increases ACh release, this effect is due to an action of atropine at the site of ACh release (Szerb, 1964). On the other hand antinicotinic agents, such as D-tubocurarine and dihydro-β-erythroidine fail to alter ACh release when applied to the cortex even in concentrations a hundred times that of atropine (Dudar and Szerb, 1969).

Various theories have been proposed to explain this effect of atropine. Acting postsynaptically, atropine could preserve ACh by occupying cholinoceptive sites which normally inactivate ACh or atropine could prevent the uptake of ACh which has been shown to occur *in vitro* (Liang and Quastel, 1969; Polak, 1969). Alternatively atropine could act presynaptically, facilitating ACh release and synthesis as proposed by Bertels–Meeuws and Polak (1968). Finally MacIntosh (1963) suggested that atropine might increase ACh release by blocking transmission in a cholinergic synapse which is a part of an inhibitory loop controlling the activity of cholinergic fibres.

Recent results (Dudar and Szerb, 1969) have shown that atropine fails to increase ACh release when the resting ACh output is low, due either to halothane–nitrous oxide anaesthesia, or to an absence of activity in the cholinergic fibres ascending to the cortex following lesions in the reticular formation or following the topical application of tetrodotoxin. This indicates that atropine increases ACh output only when ACh is being released as a result of activity in cholinergic neurones. To see whether activity of the whole cholinergic neurone or only that of the cortical terminals is necessary for the action of atropine, increased ACh release was evoked under halothane–N_2O anaesthesia before and after the application of atropine with stimulation of either the cortical surface or the reticular formation.

References pp. 164–165

Since the somata of the great majority of cortical cholinergic fibres are located subcortically (Hebb *et al.*, 1963; Krnjević and Silver, 1965), stimulation of the surface will activate only the terminals orthodromically, while stimulation of the reticular formation activates the cholinergic fibres trans-synaptically (Szerb, 1967). Atropine approximately doubled the ACh release due to surface stimulation but increased 4–5 times ACh release resulting from reticular formation stimulation (Fig. 1). These results can be interpreted as follows: for the full effect of atropine on ACh output it is necessary that the somata of cholinergic neurones be synaptically activated either

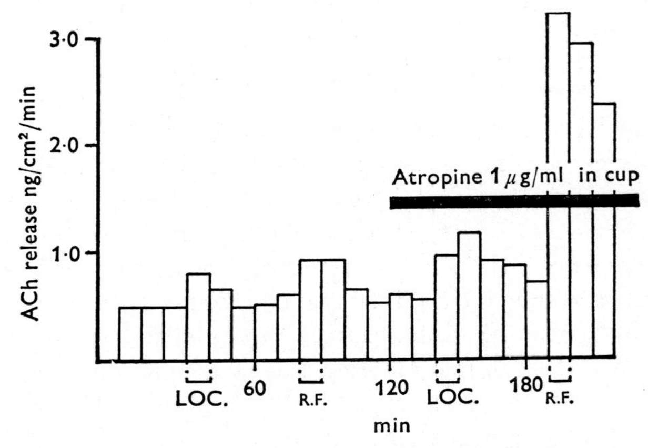

Fig. 1. Effect of atropine on ACh output from the parietal cortex of the cat under halothane–N_2O anaesthesia. Loc., bipolar stimulation of the collection area (20 V, 1 msec, 30/sec); R.F., stimulation of the reticular formation (6 V, 0.3 msec, 100/sec for 1 sec every 10 sec). (Reproduced from *J. Physiol. (Lond.)*, (1969), J. D. DUDAR AND J. C. SZERB, **203**, 741–762.)

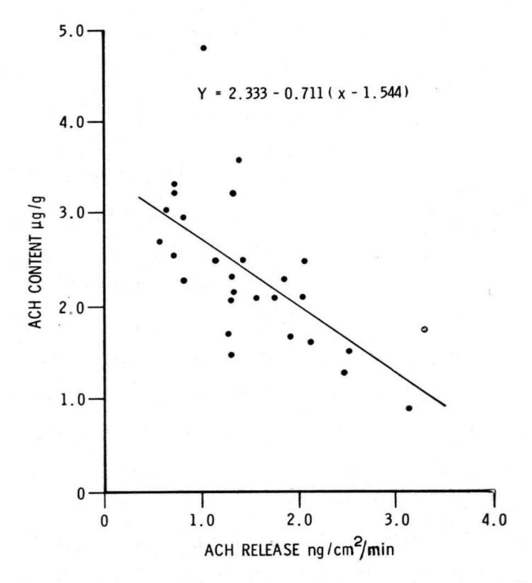

Fig. 2. Relationship between cortical ACh output and content in 28 pretrigeminally sectioned non-anaesthetised cats. Points represent the average of 4 initial values for output and of two initial values for content. (Reproduced from *Can. J. Physiol. Pharmacol.* (1970), J. C. SZERB *et al.*, **48**, 780–790.)

spontaneously or by stimulation of the reticular formation. They exclude interference with the inactivation of ACh as a cause of its increased release by atropine but are consistent with the hypothesis that atropine, by blocking transmission across a cholinergic synapse in the cortex, removes an inhibitory influence on the somata of cholinergic neurones. The smaller potentiating effect of atropine on ACh release due to stimulation of the cortical surface could be due to the facilitatory effect of atropine on ACh release as described by Bertels–Meeuws and Polak (1968), who found that atropine enhances ACh release from cortical slices incubated with 25 mM KCl.

The effect of atropine and other drugs affecting ACh release and content in the cortex has been investigated further (Szerb *et al.*, 1970) by following ACh release in one hemisphere and taking repeated small biopsies from the opposite side in pretrigeminally sectioned, non-anaesthetised cats. Without any treatment there was an inverse relationship between ACh release and content. (Fig. 2). Decreasing ACh release by pentobarbitone or by placing a lesion in the midbrain reticular formation caused an increase in content, while an increase in ACh release after intravenous

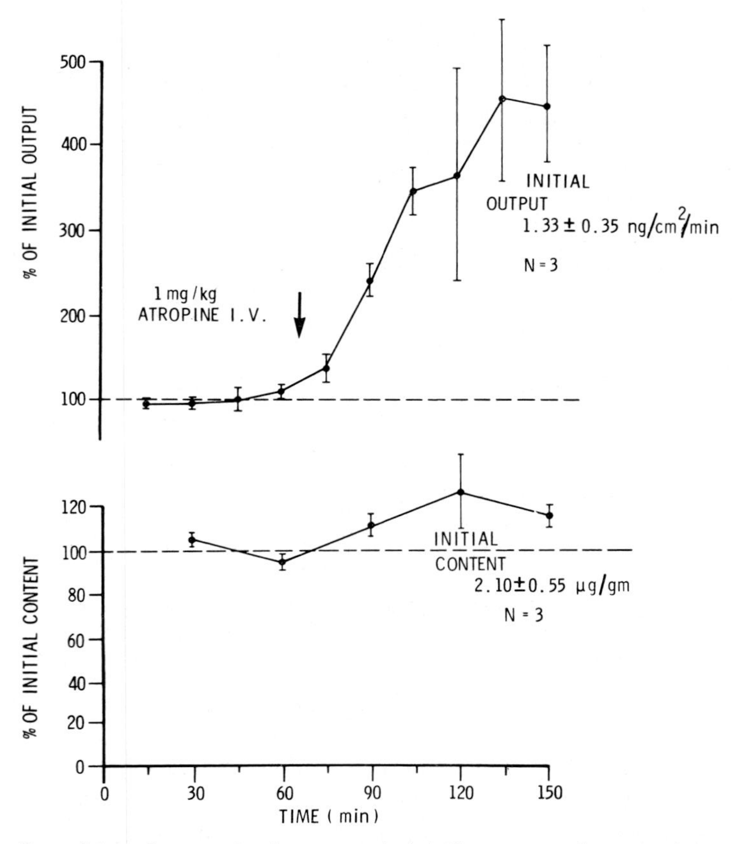

Fig. 3. The effect of 1 mg/kg atropine i.v. on cortical ACh output and content in pretrigeminally sectioned non-anaesthetised cats. (Reproduced from *Can. J. Physiol. Pharmacol.* (1970), J. C. SZERB, *et al.*, **48**, 780–790.)

picrotoxin in cats anaesthetised with Dial was accompanied by a decreased ACh content on the opposite side. Atropine (1 mg/kg i.v. or 1 μg/ml applied to the cortex) increased ACh output without decreasing ACh content (Fig. 3). Only when given in large doses (25 mg/kg i.v.) did atropine decrease the amount of ACh in the cortex. However, pretreatment with 1 mg/kg atropine greatly enhanced the effect of picrotoxin both on the release and content of ACh.

These results show that, with the exception of small doses of atropine, increased ACh release is accompanied by a decrease in ACh content. Had atropine increased ACh release solely by increasing activity in cholinergic neurones a similar decrease in content would have been found. Therefore, atropine might act also directly on the release mechanism as proposed by Bertels–Meeuws and Polak (1968).

The turnover rate of ACh was further investigated by measuring changes in ACh release and content following the local application of the ACh-synthesis inhibitor hemicholinium-3. In the absence of atropine, a fall in ACh content was not accompanied by a significant drop in release, nor was there a decrease in ACh content measured after cholinesterase inhibition (Fig. 4). However, in the presence of atropine

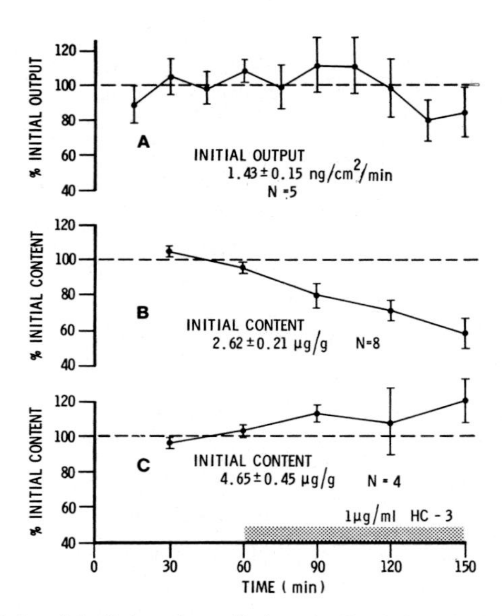

Fig. 4. Effect of 1 μg/ml hemicholinium-3 applied topically on cortical ACh output and content without or with the previous topical application of echothiophate iodide in pretrigeminally sectioned non-anaesthetised cats. A, ACh output; B, ACh content without echothiophate; and C, ACh content following the application for 30 min of 0.5 mg/ml echothiophate iodide. (Reproduced from *Can. J. Physiol. Pharmacol.* (1970), J. C. SZERB *et al.*, **48**, 780–790.)

the high rate of ACh release decreased rapidly following hemicholinium-3 application and the content measured without cholinesterase inhibition decreased faster than without atropine (Fig. 5). ACh present after cholinesterase inhibition showed a rather variable increase following hemicholinium-3. The absence of a drop in ACh release when no atropine was present in spite of a 42% decline in content can be explained

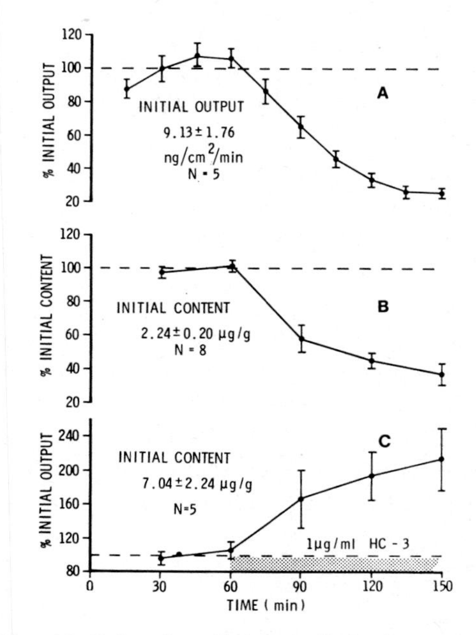

Fig. 5. Effect of 1 µg/ml hemicholinium-3 applied topically in the presence of 1 µg/ml atropine on cortical ACh output and content without or with the previous topical application of echothiophate iodide in pretrigeminally sectioned non-anaesthetised cats. A, ACh output; B, ACh content without echothiophate, and C, ACh content following the application for 30 min of 0.5 mg/ml echothiophate iodide. (Reproduced from *Can. J. Physiol. Pharmacol.* (1970), J. C. Szerb *et al.*, **48**, 780–790.)

TABLE I

THE TURNOVER RATE OF ACh IN THE CORTEX

Anaesthesia	*Drug or lesion*	*Fraction of ACh content released per min*
Pretrigeminally		
sectioned non-anaesthetised	—	$^1/_{125}$
sectioned non-anaesthetised	1 mg/kg atropine i.v.	$^1/_{33}$
sectioned non-anaesthetised	Reticular formation lesion	$^1/_{500}$
Dial anaesthesia	—	$^1/_{180}$
Dial anaesthesia	1 mg/kg atropine i.v.	$^1/_{96}$
Dial anaesthesia	2 mg/kg picrotoxin i.v.	$^1/_{40}$
Dial anaesthesia	1 mg/kg atropine + 2 mg/kg picrotoxin i.v.	$^1/_{12}$

by an increase in the activity of the cholinergic fibres whose firing rate is governed by a negative feedback loop as proposed by Dudar and Szerb (1969). In the presence of atropine this contiol system is no longer operative and output rapidly declines as the content falls. The decrease in ACh content following hemicholinium-3 is faster in the presence of atropine than without it, indicating that atropine enhances ACh release

References pp. 164–165

rather than interfering with the inactivation of ACh. ACh content measured after cholinesterase inhibition does not reflect the releasable fraction of presynaptic ACh, probably because it includes surplus ACh and ACh which has been taken up after release.

By relating ACh release and reduction in content following inhibition of ACh synthesis by hemicholinium-3 it could be shown that about one-third of the ACh released appears in the collection fluid held in the Perspex cylinder. This allowed the calculation of the turnover rate of cortical ACh when both content and release were measured. Table I shows ACh turnover rates calculated from these experiments. These rates are much slower than those calculated by Schuberth *et al.* (1970) from results based on the incorporation of labelled choline into ACh stores. However, because of some uncertainty about the size of the choline pool which is the precursor of ACh, their values cannot be considered definitive. It can be seen that although atropine in small doses does not decrease ACh content, it increases ACh turnover, and especially enhances the effect of a stimulant such as picrotoxin, on ACh turnover rate.

SUMMARY

Atropine appears to affect ACh release in the cortex in two ways:

(1) By blocking a recurrent inhibitory loop which governs the activity of ascending cholinergic fibres. This can only be shown *in vivo*.

(2) By facilitating the release and synthesis of ACh, which occurs both *in vitro* and *in vivo*.

REFERENCES

BERTELS–MEEUWS, M. M. AND POLAK, R. L. (1968) Influence of antimuscarinic substances on *in vitro* synthesis of acetylcholine by rat cerebral cortex. *Brit. J. Pharmacol.*, **33**, 368–380.

CRAWFORD, J. M. AND CURTIS, D. R. (1966) Pharmacological studies on feline Betz cells. *J. Physiol. (Lond.)*, **186**, 121–138.

DUDAR, J. C. AND SZERB, J. C. (1969) The effect of topically applied atropine on resting and evoked cortical acetylcholine release. *J. Physiol. (Lond.)*, **203**, 741–762.

HEBB, C. O., KRNJEVIĆ, K. AND SILVER, A. (1963) Effect of undercutting on the acetylcholinesterase and choline acetyltransferase activity in the cat's cerebral cortex. *Nature (Lond.)*, **198**, 692.

KRNJEVIĆ, K. AND PHILLIS, J. W. (1963) Pharmacological properties of acetylcholine-sensitive cell in the cerebral cortex. *J. Physiol. (Lond.)*, **166**, 328–350.

KRNJEVIĆ, K. AND SILVER, A. (1965) A histochemical study of cholinergic fibres in the cerebral cortex. *J. Anat. (Lond.)*, **99**, 711–759.

LIANG, C. C. AND QUASTEL, J. H. (1969) Effects of drugs on the uptake of acetylcholine in rat brain cortex slices. *Biochem. Pharmacol.*, **18**, 1187–1194.

MACINTOSH, F. C. (1963) Synthesis and storage of acetylcholine in nervous tissue. *Canad. J. Biochem.*, **41**, 2555–2571.

MITCHELL, J. F. (1963) The spontaneous and evoked release of acetylcholine from the cerebral cortex. *J. Physiol. (Lond.)*, **165**, 98–118.

POLAK, R. L. (1969) The influence of drugs on the uptake of acetylcholine by slices of rat cerebral cortex. *Brit. J. Pharmacol.*, **36**, 144–152.

SCHUBERTH, J., SPARF, B. AND SUNDWALL, A. (1970) On the turnover of acetylcholine in nerve endings of mouse brain *in vivo*. *J. Neurochem.*, **17**, 461–468.
SZERB, J. C. (1964) The effect of tertiary and quaternary atropine on cortical acetylcholine output and electroencephalogram in cat. *Can. J. Physiol. Pharmacol.*, **42**, 303–314.
SZERB, J. C. (1967) Cortical acetylcholine release and electroencephalographic arousal. *J. Physiol. (Lond.)*, **192**, 329–343.
SZERB, J. C., MALIK, H. AND HUNTER, E. G. (1970) Relationship between acetylcholine content and release in the cat's cerebral cortex. *Can. J. Physiol. Pharmacol.*, **48**, 780–790.

DISCUSSION

WALKER: In your diagram you showed an inhibitory input onto the cholinergic neurone. Is it possible to block this inhibition?

SZERB: For this we would need to know where the somata of the cortical cholinergic neurones lie and there is some uncertainty about their anatomical location. Indications are that part of them are in the septum, at least in the cat. If you ask me whether we can block their inhibitory input directly, the answer is no.

POLAK: May I ask if I understood correctly that the original finding of an inverse relationship between ACh content on the one hand and ACh release on the other does not apply to the effect of atropine as soon as you give a cholinesterase inhibitor?

SZERB: All determinations of ACh content, with the exception of a few with hemicholinium-3, were done without cholinesterase inhibition. With cholinesterase not inhibited on the side of the biopsies there was an inverse relationship between output and content, except with small doses of atropine.

POLAK: Well, this is where our results obtained with cortical slices from rat brain seem to be different from your results. We always used a cholinesterase inhibitor but our results depended on the choice of the cholinesterase inhibitor. When the organo-phosphorous compound Soman was used, stimulation of the release and synthesis of ACh by atropine was followed by an increase in the amounts of ACh which could be extracted from the tissue after incubation. This increase appeared to be caused by partial re-uptake of the relatively large amounts of ACh released under the influence of atropine. We found that physostigmine largely prevented this re-uptake With physostigmine in the medium, atropine also stimulated the release and synthesis of ACh but now these effects were accompanied by a decrease in the "total extractable" ACh content of the tissue.

KERKUT: When you get one-twelfth of the total content of ACh released a minute, how long do you think this would last?

SZERB: Not very long. I doubt very much whether there would be an exponential decline in output after ACh synthesis was inhibited. Therefore, the turnover estimates are given not in the form of half-time but as a fraction of the ACh present which is released in a minute.

BRADLEY: Would undercutting the cortex test your hypothesis for the inhibitory circuit?

SZERB: We have done that. Acute undercutting reduced the stimulating effect of atropine on ACh release.

Factors Influencing the Release of Prostaglandins from the Cerebral Cortex

GILLIAN M. R. SAMUELS*

Department of Pharmacology (Preclinical), The Medical School, Birmingham (Great Britain)

Several pharmacologically active substances are present in superfusates of the cerebral cortex, including acetylcholine, 5-hydroxytryptamine, substance P and the prostaglandins. Members of this latter family of hydroxy-unsaturated fatty acids are released from a great variety of tissues on nerve and hormone stimulation (Ramwell and Shaw 1970), and have been shown to possess a wide range of biological activity (Horton, 1969).

The release of prostaglandins from the cerebral cortex of the anaesthetised cat was first reported by Ramwell and Shaw (1966). The level of release could be increased by stimulation of the contralateral forepaw and by administration of analeptics, suggesting that prostaglandin release might be connected with neuronal activity.

Using the unanaesthetised feline *encéphale isolé* preparation it has been possible to show a correlation between the level of prostaglandin release and the spontaneous fluctuations in the neuronal activity of the cortex as monitored in the electrocorticogram (E.Co.G.). Electrocortical desynchronisation and arousal may be evoked by high-frequency electrical stimulation of the brain stem reticular activating system (Moruzzi and Magoun 1949), and is accompanied by a significant increase in prostaglandin efflux (Bradley *et al.*, 1969).

Changes in electrocortical activity and behaviour can also be induced by drug administration (Bradley and Elkes, 1957). The effects of 5 such drugs on prostaglandin output was investigated. The drugs used fall into two groups: those which affect electrocortical activity and behaviour in parallel, amphetamine, chlorpromazine and pentobarbitone, and those which cause a "pharmacological dissociation" between the E.Co.G. and behaviour, atropine and physostigmine.

Electrocortical desynchronisation whether produced by amphetamine (0.5 mg/kg) or physostigmine (1 mg total dose) was accompanied by a significant increase in prostaglandin release. Drugs which induce electrocortical synchronisation, chlorpromazine (2.5 mg/kg), pentobarbitone (10 mg/kg), and atropine (2 mg/kg), all caused a significant decrease in prostaglandin release (Bradley and Samuels, unpublished observations).

Thus, it appears that electrocortical desynchronisation, either spontaneous or

* Present address: Tunstall Laboratory, Shell Research Ltd, Sittingbourne, Kent.

References pp. 169

induced, and whether or not it was accompanied by behavioural changes could be correlated with an increase in the level of prostaglandin release from the cerebral cortex of the feline *encéphale isolé* preparation. Conversely drug-induced synchronisation was accompanied by a decrease in prostaglandin release. The level of prostaglandin release from the cerebral cortex therefore appeared to be more closely correlated with the pattern of electrocortical activity than with the level of behavioural arousal.

REFERENCES

Bradley, P. B. and Elkes, J. (1957) The effect of some drugs on the electrical activity of the brain. *Brain*, **80**, 77–117.

Bradley, P. B., Samuels, G. M. R. and Shaw, J. E. (1969) Correlation of prostaglandin release from the cerebral cortex of cats with the electrocorticogram following stimulation of the reticular formation. *Brit. J. Pharmacol.*, **37**, 151–157.

Holmes, S. W. and Horton, E. W. (1968) The identification of four prostaglandins in dog brain and their regional distribution in the central nervous system. *J. Physiol. (Lond.)*, **195**, 731–741.

Horton, E. W. (1969) Hypotheses on the physiological roles of prostaglandins. *Physiol. Rev.*, **49**, 122–161.

Moruzzi, G. and Magoun, H. W., (1949) Brain stem reticular formation and activation of the E.E.G. *Electroenceph. clin. Neurophysiol.*, **1**, 455–473.

Ramwell, P. W. and Shaw, J. E. (1966) Spontaneous and evoked release of prostaglandins from the cerebral cortex of anaesthetised cats. *Amer. J. Physiol.*, **211**, 125–134.

Ramwell, P. W. and Shaw, J. E. (1970) Biological significance of the prostaglandins. *Recent Prog. Horm. Res.*. **26**, 139–187.

DISCUSSION

Rick: It might be interesting in terms of the drugs which are affecting levels of arousal to look at some drugs which do not decrease the level of glucose metabolism in brain tissue while they do affect the level of arousal, for example γ-hydroxybutyrate. Such drugs, instead of producing a state of anaesthesia, tend to produce a state which is very akin to sleep and the levels of prostaglandins may give a very interesting indication as to whether there were any differences from what you obtain with, for example, pentobarbitone.

Szerb: What is the significance of these findings?

Samuels: The significance of the prostaglandins in the CNS is obscure but peripheral studies make one suspect that they could be implicated as modulators of transmission as part of feed back systems. This is particularly interesting in the CNS when one considers the fine control that is required. The structure of the prostaglandins and their particular physico-chemical properties (they are highly hydrophilic and lipophilic) means that they are ideally suited for a role as membrane perturbers, and, as such, could be implicated in differential permeability of membranes to various ions. Also, there is another interesting aspect in that they are vasoactive compounds and there is a very low level of prostaglandin dehydrogenase which is responsible for biological deactivation of prostaglandins in the brain, so it is possible that they may have another role in the brain as local regulators of blood flow. One could envisage release from an activated tissue causing increase in blood flow to that tissue.

Fonnum: Do the prostaglandins show any variation in regional distribution?

Samuels: Holmes and Horton (1968) have been unable to show regional distribution in dog brain. As far as I am aware this is the only species in which regional distribution has been studied. Kataoka *et al.* (1967) have shown in rats a certain localization in non-cholinergic nerve endings, but obviously this is not the only place where it is found (Hopkin *et al.* 1968.)

FONNUM: Are they bound to membranes?

SAMUELS: As far as I know there is no evidence that they are membrane bound, although they may be released from the membrane.

MOLENAAR: Could there be any release of prostaglandin from blood vessels in the cortex?

SAMUELS: This is a possibility, but even if part of the release we are measuring is coming from blood vessels I do not think it is a significant proportion. On administration of amphetamine to an animal there is an increase in BP and then there is an increase in release of some prostaglandin A compounds. These are highly vasoactive compounds and may come from the walls of blood vessels.

WALKER: What is the evidence for prostaglandins acting as modulators of synaptic activity?

SAMUELS: I don't think there is any direct evidence, but prostaglandins may play a part in synaptic transmission in the cerebellum, where it has been shown that they antagonise the effect of noradrenaline on Purkinje cells.

REFERENCES

HOLMES, S. W. AND HORTON, E. W. (1968) The identification of four prostaglandins in dog brain and their regional distribution in the central nervous system. *J. Physiol. (Lond.)*, **195**, 731–741.

HOPKIN, J. M., HORTON, E. W. AND WHITTAKER, V. P. (1968) Prostaglandin content of particulate and supernatant fractions of rabbit brain homogenates, *Nature, (Lond.)*, **217**, 71–72.

KATAOKA, K., RAMWELL, P. W. AND JESSUP, S. (1967) Prostaglandins: Localisation in subcellular particles of rat cerebral cortex. *Science*, **157**, 1187–1189.